A compact fast-neutron producing target for high resolution cross section measurements

Cover illustration: Top view of the GELINA facility

The research described in this thesis was performed as a cooperation of the following institutions:
Department of Radiation, Radionuclides & Reactors of the Faculty of Applied Sciences, Delft University of Technology, Mekelweg 15, 2629 JB Delft, The Netherlands,
Neutron Physics Unit of the Institute for Reference Materials and Measurements (IRMM), Joint Research Centre (JRC), European Commission (EC), Retieseweg 111, B-2440 Geel, Belgium.
Department of Multi-Scale Physics of the Faculty of Applied Sciences, Delft University of Technology, Prins Bernhardlaan 6, 2628 BW Delft, The Netherlands,

The author would like to thank EC-JRC-IRMM for the financial support.

A compact fast-neutron producing target for high resolution cross section measurements

Proefschrift

ter verkrijging van de graad van doctor
aan de Technische Universiteit Delft,
op gezag van de Rector Magnificus prof. dr. ir. J.T. Fokkema,
voorzitter van het College voor Promoties,
in het openbaar te verdedigen

op maandag 24 april 2006 om 10:30 uur

door

Marek FLAŠKA

inžinier,
Slovenská Technická Univerzita Bratislava
geboren te Veľký Krtíš, Slowakije

Dit proefschrift is goedgekeurd door de promotoren:

Prof. em. dr. ir. H. van Dam
Prof. dr. ir. T.H.J.J. van der Hagen
Prof. dr. ir. C.R. Kleijn

Samenstelling promotiecommissie:

Rector Magnificus,	voorzitter
Prof. em. dr. ir. H. van Dam,	Technische Universiteit Delft, promotor
Prof. dr. ir. T.H.J.J. van der Hagen,	Technische Universiteit Delft, promotor
Prof. dr. ir. C.R. Kleijn,	Technische Universiteit Delft, promotor
Prof. em. dr. H. Postma,	Technische Universiteit Delft
Prof. dr. ir. Th.H. van der Meer,	Universiteit Twente
Prof. dr. E. Grosse,	Technische Universität Dresden
Prof. dr. P. Rullhusen,	Universität Göttingen

ISBN: 1-58603-610-6

Published by IOS Press under the imprint Delft University Press

IOS Press
Nieuwe Hemweg 6b
1013 BG Amsterdam
The Netherlands
fax: +31-20-687 0019
email: order@iospress.nl

LEGAL NOTICE
The publisher is not responsible for the use which might be made of the following information.

PRINTED IN THE NETHERLANDS

Aan Csaba

Table of contents

List of symbols   x

List of abbreviations   xiii

Chapter 1:   Introduction   **1**
  1.1   Prologue   1
  1.2   The time-of-flight method   2
  1.3   Aims and context of the present work   5

Chapter 2:   The GELINA time-of-flight facility   **9**
  2.1   Linear electron accelerator   10
    2.1.1   Electron gun   11
    2.1.2   Accelerator sections and klystron modulators   12
  2.2   Compression magnet   13
  2.3   The GELINA neutron producing target   16
  2.4   GELINA flight paths   17

Chapter 3:   Monte Carlo simulations for the GELINA rotary target   **19**
  3.1   GELINA neutron source modeling   20
    3.1.1   MCNP4C3 model geometry   20
    3.1.2   Photonuclear data   21
      3.1.2.1   Photoneutron production   21
      3.1.2.2   MCNP4C3 data libraries   23
    3.1.3   Variance reduction techniques   23
    3.1.4   Other important aspects of the target modeling   25
    3.1.5   Implementation of the delay distance concept in MCNP4C3   25
  3.2   Benchmarking of the MCNP model – absolute fluxes   25
    3.2.1   The moderated neutron flux   26
      3.2.1.1   Experimental setup   26
      3.2.1.2   Results and discussion   27
    3.2.2   The unmoderated neutron flux   28
      3.2.2.1   Experimental setup   28
      3.2.2.2   Results and discussion   30
  3.3   Benchmarking of the MCNP model – resolution functions   31
    3.3.1   Resolution functions of the moderated neutron spectrum   31
    3.3.2   Resolution function of the direct neutron spectrum   34
    3.3.3   Comparison of the resolution functions for different angles   36

3.3.4   Comparison of the calculated and measured resolution functions 39
3.4   Conclusions 41

Chapter 4: Influence of the target properties on the neutron production and the resolution functions **43**
4.1   Optimization of the target size – neutron-yield point of view 43
   4.1.1   Particle energy loss in medium 43
   4.1.2   Electromagnetic cascade 45
   4.1.3   Target features affecting neutron production 46
4.2   Optimization of the target size – resolution-function point of view 48
   4.2.1   Target features affecting resolution functions – neutron high-energy range 48
   4.2.2   Target features affecting resolution functions – neutron low-energy range 51
   4.2.3   The resolution functions of the compact target using the existing moderator 53
4.3   Conclusions - What is the way to optimize a new target? 55

Chapter 5: Towards a compact neutron producing target – neutronics point of view **57**
5.1   The neutron flux 57
   5.1.1   The choice of the target material for a new compact target 57
   5.1.2   Angle-dependent flux 58
   5.1.3   Compact U-Mo neutron producing targets 60
5.2   Resolution functions 63
   5.2.1   The comparison of alternative target materials 63
   5.2.2   Compact U-Mo neutron producing targets 66
5.3   Figure of merit 67
   5.3.1   The comparison of alternative target materials 69
   5.3.2   Compact U-Mo neutron producing targets 71
5.4   Conclusions 73

Chapter 6: Towards a compact neutron producing target – heat transfer point of view **75**
6.1   Initial target optimization 76
   6.1.1   Segmentation of the compact U-Mo design 77
   6.1.2   Plate by plate optimization 78
6.2   Computational fluid dynamics simulations with the FLUENT code 82
   6.2.1   FLUENT code 82
   6.2.2   GAMBIT preprocessor 83

6.2.3 Accuracy of the computational fluid dynamics simulations 83
6.2.3.1 Accuracy and suitability of applied flow models 84
6.2.3.2 Numerical accuracy of computational fluid dynamics simulations 85
6.3 Further target optimization using coupled FLUENT and MCNP 86
6.3.1 MCNP heat source and interpolation to Fluent 86
6.3.2 Assumptions applied in the FLUENT optimization 87
6.4 Initial target design – results and discussion 88
6.5 Final target design 94
6.5.1 CFD model settings 95
6.5.2 Accuracy and grid dependence of the results 96
6.5.3 Results and discussion 99
6.6 Conclusions 107

Chapter 7: Final conclusions **109**

Summary 113

Samenvatting 117

References 121

List of publications 125

Acknowledgements 127

Curriculum vitae 129

List of symbols

A	atomic mass
A_s	side area of a plate
c	speed of light
c_p	specific heat
d	delay distance
D	half thickness of a plate
E	initial electron energy
E_C	critical energy
E_{center}	energy centroid
E_n	neutron energy
E_S	classical constant of multiple scattering theory
$FWHM_L$	full-width-half-maximum value of a resolution function
$FWHM_t$	full-width-half-maximum value of an electron pulse
h	heat transfer coefficient
H	cylinder height
I	mean ionization energy of a medium atom
I_n	neutron source intensity
l_i	distance traveled by neutron between collisions $i\text{-}1$ and i
l_n	distance between collisions $n\text{-}1$ and n
L	flight path length
L_{char}	characteristic length of a domain
L_{ref}	reference source-detector distance
L_{sim}	detector distance in a simulation
m_e	electron mass
m_n	neutron mass
N_A	Avogadro's number
N_U	Nusselt number
q	heat generation rate in one half of a plate
q_c	particle charge
q^*	modified heat generation rate
q'''	heat generation rate per unit of volume
Q	total power
r	radial coordinate
r_e	electron radius

R	cylinder radius
Re	Reynolds number
Re_{crit}	critical Reynolds number
R_M	Moliere radius
S	channel cross section
t	time of flight
tme	time variable
T	moderator thickness
$T_{channel}$	channel temperature
T_{inlet}	inlet temperature
T_{max}	maximum temperature
T_{wall}	wall temperature
$T(r)$	radius-dependent temperature
$T(x)$	x-coordinate-dependent temperature
T_0	surface temperature
v	velocity
vel	velocity variable
v_{Hg}	mercury velocity
v_i	neutron velocity between collisions *i-1* and *i*
v_{max}	maximum velocity
v_n	neutron velocity between collisions *n-1* and *n*
V	volume
x, xxx	x-coordinate variable
x	areal density
X	channel half thickness
X_0	radiation length
y, yyy	y-coordinate variable
y^+, y^*	distance in wall units
z, zzz	z-coordinate variable
Z	atomic number
$4\pi Y_n$	neutron production integrated over 4π
β	ratio *v/c*
$\delta E_n / E_n$	relative energy resolution
δL	uncertainty of flight path length
δt	time uncertainty
Δp	pressure drop
ΔT	temperature drop
Δl_n	distance traveled by neutron in moderator after last collision

ε_0	vacuum permittivity
Φ	neutron flux
Φ_m	mass flow rate
Γ_{tot}	FWHM of a natural resonance
λ	thermal conductivity
λ_{Hg}	thermal conductivity of mercury
μ	dynamic viscosity
μ_t	turbulent viscosity
ν	number of neutrons per reaction
θ	angle
ρ	density
ρ_{Hg}	mercury density
σ^{prod}	neutron production cross section
$(dE/dx)_{ion}$	ionization energy loss of a charged particle
$(dE/dx)_{rad}$	electron energy loss due to bremsstrahlung
$d\phi_{MCNP}/d\ln E$	MCNP neutron flux per unit lethargy

ACE	nuclear data format
CFD	Computational Fluid Dynamics
EC	European Commission
ENDF	Evaluated Nuclear Data File
FOM	Figure Of Merit
FWHM	Full Width at Half of the Maximum
FWHUM	Full Width at one HUndredth of the Maximum
FWTM	Full Width at one Tenth of the Maximum
GDR	Giant Dipole Resonance
GELINA	Geel Electron LINear Accelerator
HF	High Frequency
IAEA	International Atomic Energy Agency
IRMM	Institute for Reference Materials and Measurements
JENDL	Japanese Evaluated Nuclear Data Library
JRC	Joint Research Centre
MCNP	Monte Carlo N-Particle code
mfp	mean free path
NJOY	nuclear data processing system
PD	Probability Density
REFIT_IRMM	general REsonance FITting program modified at IRMM
RF	Resolution Function
TOF	Time Of Flight
UDF	User-Defined Function
UDM	User-Defined Memory
U-Mo	Uranium-Molybdenum alloy

Chapter 1

Introduction

1.1 Prologue

The development and improvement of a comprehensive neutron cross section database is essential for many areas of research and technology. Numerous interaction types may occur in various isotopes over a broad energy area. By far the most important interactions relevant for nuclear power production are the neutron-induced reactions.

The probability for a certain neutron interaction is proportional to the neutron flux and to the isotopic density of the material under consideration. The proportionality constant is called the neutron cross section. A proper knowledge of neutron cross sections is of a great importance when evaluating the safety and risk related to operation of nuclear power plants, nuclear waste management, accelerator-driven transmutation systems or new concepts of nuclear power production. Reducing uncertainties in the neutron cross section data can result in an enhanced safety of present and future nuclear power systems. Accurate neutron cross sections also play a crucial role in many other disciplines such as astrophysics, medicine, and security [Sal02, Smi02, Qai02, Bir03].

To perform accurate neutron cross section measurements, three energy domains need to be distinguished:
- the resolved resonance region where the neutron cross sections reveal a complicated resonance structure. Here, experiments with a very high resolution in energy are required. The use of pulsed particle beams using accelerators, combined with a **Time-Of-Flight** (TOF) technique, is the only possibility to perform measurements at the required resolution. Dedicated TOF facilities make use of such accelerators in combination with one or several neutron flight paths,

- the unresolved resonance region, in which the experimental resolution does not allow to separate the resonances. Also here the TOF facilities are the best choice,
- very high energies, e.g. above the thresholds for the production of charged particles. In this energy region measurements do not require the high resolution needed in the resonance region. Here also measurements with mono-energetic neutrons may be performed, which are usually obtained using nuclear reactions produced with low-energy hadron accelerators.

In the energy interval from thermal neutron energies to a few MeV the cross sections have a resonance-type energy dependence and large differences exist between the neighboring isotopes. Accurate resonance parameters, that describe the cross sections in detail, are essential for a proper account of reaction rates and the detailed neutron flux distribution in many applications. In particular, the effect of resonance self-shielding may considerably influence reaction rates [Sta01]. Therefore high-resolution measurements are needed to allow extraction of the resonance parameters, by using the technique of resonance shape analysis [Gun00, Cor02]. The required measurement accuracy can only be obtained at TOF facilities specially designed for a very high resolution in energy.

Among the neutron TOF facilities available in the world, the **Ge**el **E**lectron **LIN**ear **A**ccelerator (GELINA) facility [Ben78] of the **J**oint **R**esearch **C**entre (JRC) of the **E**uropean **C**ommission (EC) is the one with the best energy resolution. It is the objective of this work to investigate the possibilities to improve even further the present capabilities of this neutron data measurement facility. It will be shown that major improvements can only be obtained by designing a new high-power neutron producing target. The new target designed in the framework of this thesis will substantially enhance the obtainable energy resolution, especially in the energy region above 100 keV, while not compromising the presently available neutron flux.

1.2   The time-of-flight method

In a TOF facility, the neutrons used for the neutron cross section measurements are produced by the impact of a short pulse of high-energy particles on a neutron producing target. The impinging particles can be:
- electrons that create neutrons via the production of bremsstrahlung and consecutive photonuclear reactions,
- protons that generate neutrons via the spallation reaction.

The neutrons travel along a flight path towards the experiment. The energy E_n of the neutron is measured by taking the time difference t between the impact of the impinging particle on the neutron target and the subsequent detection of the neutron at the experiment located at a given flight path distance L. The energy of the neutron can be calculated from the nonrelativistic expression

$$E_n = \frac{1}{2} m_n \frac{L^2}{t^2} = \left(\frac{72.3L}{t} \right)^2 \tag{1.1}$$

where m_n is the neutron mass. E_n, L, and t are expressed in eV, m, and µs, or alternatively in MeV, m, and ns.

In principle there should be a one-to-one correspondence between the neutron TOF and its kinetic energy. However, even neutrons of the same final energy, which are produced as a consequence of a single pulse with negligible width, arrive at the experimental setup with a distribution in time. This is due to uncertainties on the effective length of the path traveled by the neutrons, and due to time uncertainties.

As a result, the relative energy resolution of a TOF facility is given by

$$\frac{\delta E_n}{E_n} = 2 \left(\sqrt{\left[\frac{\delta L}{L} \right]^2 + \left[\frac{\delta t}{t} \right]^2} \right) \tag{1.2}$$

where δL is the uncertainty of the effective flight path length L and δt is the uncertainty of the time t.

As shown in Equation (1.2), the total energy resolution of a TOF measurement is determined by the convolution of two independent components. Generally, uncertainties in these components (δL, δt) can be considered either as standard uncertainties conventional in error analysis or as **Full-Width-at-Half-of-the-Maximum (FWHM)** values. In this work, all uncertainties are expressed in FWHM, unless stated otherwise.

By expressing t from Equation (1.1) and inserting into Equation (1.2) the following formula is obtained:

$$\frac{\delta E_n}{E_n} = \frac{2}{L}\sqrt{\delta L^2 + 1.9E_n\delta t^2} \qquad (1.3)$$

where E_n is given in eV, L and δL in cm, and δt in µs. Equation (1.3) shows the relation between the relative energy resolution of a TOF facility and the distance and time measurement uncertainties [Coc02].

The neutron energy uncertainty consists of three major components:
- the accelerator–related component, which is given by the pulse width of the initial particles impinging on the neutron producing target,
- the target-related component, which is influenced by neutron scattering in the target,
- the component introduced by the data acquisition system and detector. This component is related to a particular experimental setup.

The distribution of the neutrons of the same energy, arriving to the detector, appears mainly due to the scattering process in a target material and the size of the neutron source. This distribution is referred to as the so-called **Resolution Function (RF)** of the target [Coc83, Bru02]. The RF is energy and angle dependent. In the resonance shape analysis of most nuclides there is a wide energy range where the apparent width of the measured resonances is affected, or even dominated, by the RF. Therefore, a good knowledge and, if possible, an improvement of these RFs is of great importance in a TOF facility.

The TOF difference can be associated with a distance difference, the so-called delay distance [Coc83, Coc96]. Using the delay distance d instead of time of flight t is a convenient way to view and parameterize the RF. The delay distance is given by

$$d = v_n t - L_{ref} \qquad (1.4)$$

where v_n is the velocity of the neutron at the detector and L_{ref} is the source-detector distance. The delay distance represents an effective flight path change, e.g. an elongation or a contraction of the flight path. The usefulness of this concept comes from the observation that the RF is a very weak function of the neutron energy when expressed in terms of delay distance, rather than a function of time. This result was first established by [Gro47] for a homogenous hydrogenous medium. This convention allows users to apply broad neutron energy bins when using Monte Carlo simulation techniques.

1.3   Aims and context of the present work

The GELINA facility is located in Geel, Belgium at the IRMM-JRC-EC site. This facility is a powerful pulsed white spectrum neutron source specially designed for neutron cross section measurements with a high resolution in energy. The TOF method is used to deduce the energy of the neutrons. At GELINA short bursts of neutrons are generated in a neutron producing target by photonuclear reactions induced by high-energy gamma radiation, the so-called bremsstrahlung. High-energy electrons from a linear accelerator previously produce this bremsstrahlung. A water moderator is used to decrease the energy of neutrons for the measurements requiring low energies. The measurements can be carried out in the energy range from 1 meV to 20 MeV.

There has been a continuous effort at GELINA to further improve its characteristics. As a result, the accelerator-related improvements have been 'exhausted' for GELINA. The recent advent of new spallation-neutron sources renewed an interest in the potential for a possible additional improvement of the GELINA facility. Therefore, a project has been launched with the main objective to investigate the potential for further enhancement of the GELINA properties. The project is presented comprehensively in this work. The objective of the project will be accomplished by building a new target with the aid of a Monte Carlo technique to study the neutron transport. Since a high power density is present in the target during the operation (10 kW), a new design must be able to safely remove the heat produced in the target material under all circumstances. This particular objective has been very challenging as the new target was predicted to be of a much smaller size than the present one. The need of having the small-size target is explained later in detail. The new target should not compromise the neutron production, while producing narrower neutron pulses. In addition, because it is also required to have a possibility to use either fast or moderated neutron spectrum for the measurements, a coolant chosen for a new target must not significantly moderate neutrons produced in the target.

<u>To summarize the main project objectives, the goal of this project is to design a new target, which would:</u>
- *minimize the RF widths in the range of high neutron energies,*
- *preserve or possibly enhance the neutron flux in the flight paths.*

The project consists of two parts, which are closely related. The first part, concerning the neutronics properties of a new target design, has been treated by the MCNP code [Brie00], which is very suitable for coupled-particle simulations. Following a preliminary investigation [Fla03, Fla04], a decision has been made to focus, within the framework of this thesis, on the optimization of the neutron target in the high-energy range. Moderator optimization will be the subject of subsequent work. For the second part, which is related to the heat removal problem, the heat transfer and fluid dynamics code FLUENT [Flu01] has been applied. Both these simulation tools are described in detail in the next chapters.

It is also important to stress that only the contributions of the GELINA accelerator and the target-moderator assembly to the energy resolution are considered. The contribution of detection system has been neglected. Consequently, the only contribution to δL from Equation 1.3 is the moderation distance in the target and moderator. The contribution to δt, on the other hand, is fully given by the width of the electron pulse.

Besides the neutron energy resolution the available neutron flux also has a significant influence on the experimental accuracy. As explained earlier, an improvement of the energy resolution while maintaining good neutron source strength has been a continuing effort at GELINA. As can be seen in the following chapter, the improvements in the past mainly concentrated on accelerator-related aspects. In essence, they have had a major influence on the δt component of the energy resolution.

All research details are described in the following chapters. In the next chapter a comprehensive overview of the existing facility is given, with emphasis on those aspects that optimize present neutron flux and energy resolution. In chapter 3 benchmark calculations for the existing target are described, which were performed to verify the Monte Carlo simulation approach. Flux calculations are shown and compared with the measurements at different flight paths. In addition, RFs are compared with earlier calculations. Further, the MCNP calculations are verified against the RFs measured at two different resonance energies for the ^{56}Fe(n,γ) reaction. Chapter 4 contains an explanation of important physical phenomena involved in the neutronics problem. Influence of the target properties on the neutron production and the RFs is the major topic of this chapter. The most relevant aspects, which have to be taken into account to optimize a new neutron producing target, are also discussed in chapter 4. In chapter 5 the neutron production and the RFs of various materials are compared to find the optimal target material. It is proven that, from the neutronics point of view, U-Mo alloy is the optimal material for a compact target to maximize the neutron production. Therefore, a similar comparison, but for different compact U-Mo designs is shown afterwards and the final target design is introduced. To judge a quality of different designs, a figure of merit is defined and evaluated. In chapter 6, an explanation is given of how the heat transfer problem has been solved for a compact U-Mo design. A combination of analytical heat transfer calculations and FLUENT simulations [Flu01] was used to optimize the geometry with respect to heat transfer. Final conclusions are given in chapter 7.

Chapter 2

The GELINA time-of-flight facility

An electron accelerator facility designed and optimized for high-resolution neutron cross section measurements must combine several features:
- the electron accelerator must produce high-power electron beam pulses. In the relevant electron energy range the neutron flux is proportional to the electron beam power. The power of the beam is equivalent to the electron energy multiplied by the electron intensity,
- the electron beam pulses must be delivered to the neutron target with the shortest possible pulse lengths so that the neutron creation is well defined in time,
- the electron pulses must be produced with a time pattern (repetition frequency, pulse spacing) that is adaptable to the experimental needs,
- the neutron producing target must be optimized for high-resolution TOF studies in the selected energy range,
- the target must also be designed to withstand the high power of the electron beam and should have a high electron-neutron conversion ratio,
- the facility must be equipped with one or preferentially with several long-distance neutron flight paths.

At the GELINA facility a unique combination of four specially designed and distinct units has been realized:
- a linear electron accelerator delivering a high-power pulsed electron beam which parameters are listed in Table 2.1,
- a post-acceleration relativistic-energy compression magnet system reducing the electron pulse width to approximately 1 ns (FWHM) while preserving the average current [Tro85],

- a rotary mercury-cooled uranium target delivering an average neutron intensity of 3.5 x 10^{13} neutrons/s, with moderator tanks placed above and below,
- 18 neutron flight paths starting in radial direction from the uranium target, leading to experimental stations located at distances between 8 and 400 m.

ELECTRON BEAM					
Pulse Length (ns)	Repetition Rate (Hz)	Peak Current (A)	Mean Current (µA)	Average Energy (MeV)	Maximum Power (kW)
without compression					
10	800	12	96	105	10
with compression					
1	800	120	96	105	10

Table 2.1. General parameters of the GELINA accelerator.

2.1   Linear electron accelerator

Figure 2.1 shows the accelerator section of GELINA. The facility has been designed to produce high-peak currents in very short pulses, with a pulse length of 10 ns, and at a repetition rate of up to 800 Hz. The energy of an electron beam leaving the last accelerator section ranges from 70 to 140 MeV.

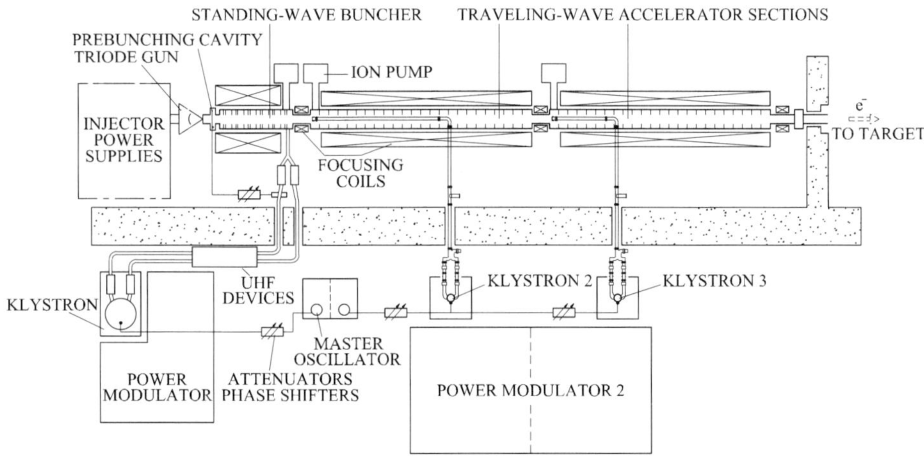

Figure 2.1. The scheme of the GELINA accelerator.

2.1.1   *Electron gun*

In order to reach the required high intensity of the beam pulse, with the prescribed time structure, the GELINA electron injector has a distinctive configuration, specially designed for this purpose. It delivers peak currents of 20 A, with a typical pulse length of 10 ns at a maximum repetition rate of 800 Hz. The electron gun is a Pierce-type triode gun with a cathode and a grid. The cathode has a negative voltage of 80 kV with respect to the anode placed at the entrance of the first accelerator section. This cathode with a diameter of 35 mm is capable of an emission of 10 A/cm² when operated in a vacuum with a pressure of 1.33 x 10⁻⁷ Pa and at a temperature of 1100 °C [Ben78]. Between the pulses the gun is cut off by the grid that is polarised at a negative voltage of 700 V referred to the cathode. To generate an intense electron beam pulse, a fast-pulse modulator delivers a positive pulse to the grid. This pulse has a nearly rectangular shape with a length of 10 ns. Its amplitude, adjustable in the range from 0 to 5 kV, defines the peak intensity of the electron pulse injected into the first accelerator section.

Before electrons enter the first section they must be bunched in short micro-pulses (bunches) at the frequency of the electromagnetic waves in the section. If not, a large part of the electrons would be lost because they would meet the electromagnetic wave in the first section during its decelerating phase. The electron pulse of 10 ns is composed of a sequence of 30 micro-pulses, with a time width of about 10 ps each. These bunches are separated by a time period of 333 ps, which corresponds to the S-band frequency of 3 GHz [Tro85].

2.1.2   Accelerator sections and klystron modulators

The GELINA accelerator consists of three S-band accelerator sections operating at 3 GHz. The electrons are accelerated along the axis of the sections by the longitudinal electric field of an electromagnetic wave traveling synchronously with the electrons. The peak amplitude of the electric field reaches 100 kV/cm. To produce this high accelerating field a peak **High-Frequency** (HF) power of 25 MW at 3 GHz is needed for each section. This pulsed HF power is delivered by three S-band klystrons. Klystrons and accelerator sections are interconnected with a waveguide network (see Figure 2.1).

To deliver a high-peak HF power, a pulsed klystron has to be driven with a high-voltage, high-current pulse. The high-power pulses with the required time structure are produced in pulse modulators. These modulators generate rectangular pulses with a peak voltage of 250 kV, a peak current of 250 A and a pulse length of 2.5 µs at a maximum repetition rate of 800 Hz. Due to extreme operating conditions (peak and average power level, high-voltage conditions, pulse repetition frequency) klystrons and pulse modulators are the most critical components of the GELINA facility.

The first accelerator section is a 2.2-m long standing-wave buncher, while the two other sections are of the traveling-wave type, with a length of 6 m each [Ben78]. Traveling-wave sections are the common choice for high-energy electron accelerators. The HF power is supplied at the beginning of the section. However, for the first GELINA section a rather unusual choice of stationary electromagnetic waves had to be made. This is needed because of the very high peak intensity of the electron pulses to be accelerated. Each electron bunch has a high charge concentration, associated with repulsive space-charge forces, as long as the electrons are not relativistic. To counteract these radial space-charge forces, a strong cylindrical symmetric solenoid magnet is required. This is incompatible with the HF power entering the accelerator section from the front side. In the first GELINA section HF power is entering the section from the backside. In this way, stationary waves, composed of forward and backward waves, are generated. However, only the forward waves are used for acceleration.

An important aspect of the operation of the GELINA facility is the filling time of the accelerator sections. The HF power is propagating in a section at a velocity of $0.2c$. The time required to fill the section with electromagnetic energy is 1.1 µs. A section must be completely filled with HF power before the electrons can be injected. This explains why the pulses produced by the klystrons must be at least 1.2-µs long. Because the electron pulse with a width of 10 ns is much shorter than the filling time, the energy gain of the accelerated electrons is proportional to the electromagnetic power that is already stored at the beginning of the pulse in the cavities of the sections. The first electron bunch 'sees' the maximum acceleration field. The next one gets lower acceleration, since part of the stored energy was already consumed by the forerunner. The charge per bunch is such that the beam loading of each bunch significantly depletes the accelerating field in the accelerator section, resulting in a linear decrease of the energy during the 10 ns-pulse from 140 to 70 MeV. This intrinsic feature of time-energy relationship, induced by beam loading, can be used for pulse compression. A relativistic electron pulse compression magnet system has been installed between the end of the third accelerator section and the neutron producing target. This system is unique in the world.

2.2   Compression magnet

The GELINA compression magnet has dimensions of 3.4 m x 3.1 m x 0.66 m and is placed horizontally in the target hall (see Figure 2.2). It consists of five sectors, which together form a 360° bending magnet. The average magnetic field produced by the magnet is 0.37 T. The curvature of a trajectory of an electron in the bending magnet is directly related to its energy. Therefore, more energetic electrons follow a longer trajectory in the magnet than those with lower energy (see Figure 2.3). Since all electrons after acceleration have a speed close to c and their energy is gradually decreasing during the 10-ns pulse, the magnet was designed in such a way that a compressed pulse appears at its exit. As a result, a 10-ns electron pulse is compressed to a 1-ns pulse. Because the electron transfer through the compression system is almost lossless, the peak amplitude of the electron pulse is increased to 120 A. A photo from the GELINA target hall is presented in Figure 2.4, in which the position of the compression magnet can be seen with respect to the rotary target.

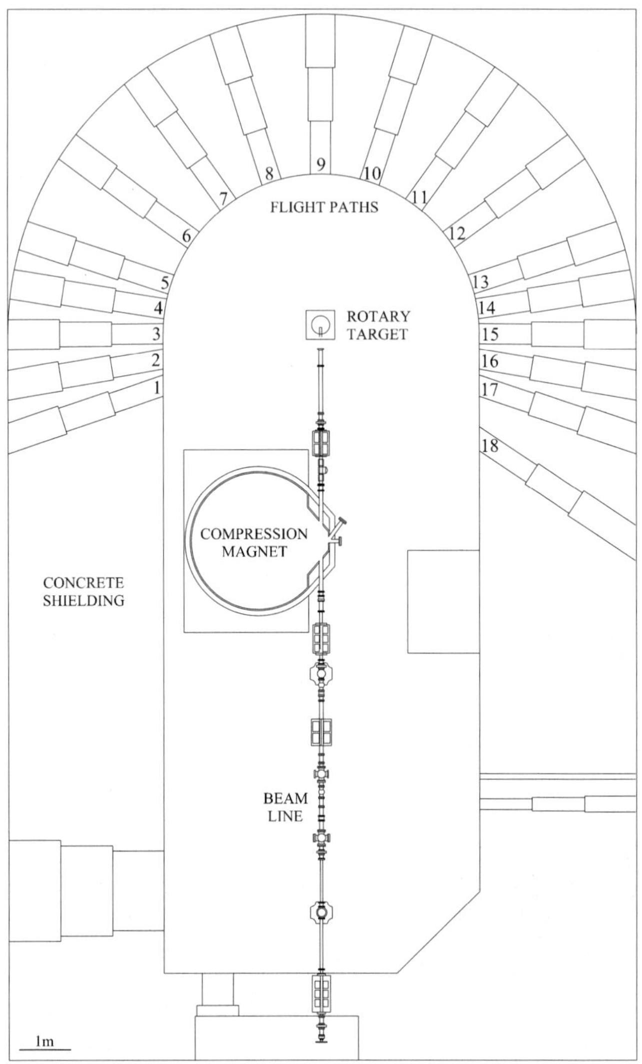

Figure 2.2. GELINA target hall with electron beam line, compression magnet and target. 18 flight paths are available (12 flight paths are presently in use), located in radial direction from the target.

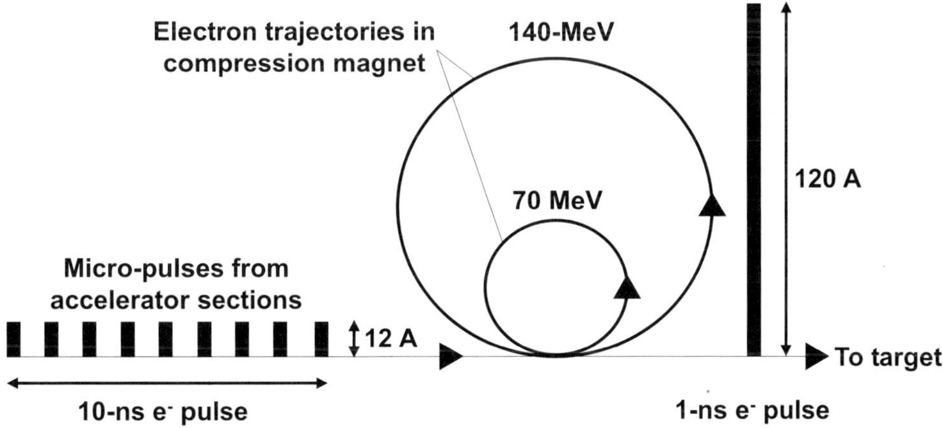

Figure 2.3. The principle of the GELINA compression magnet. A compression factor of 10 is reached.

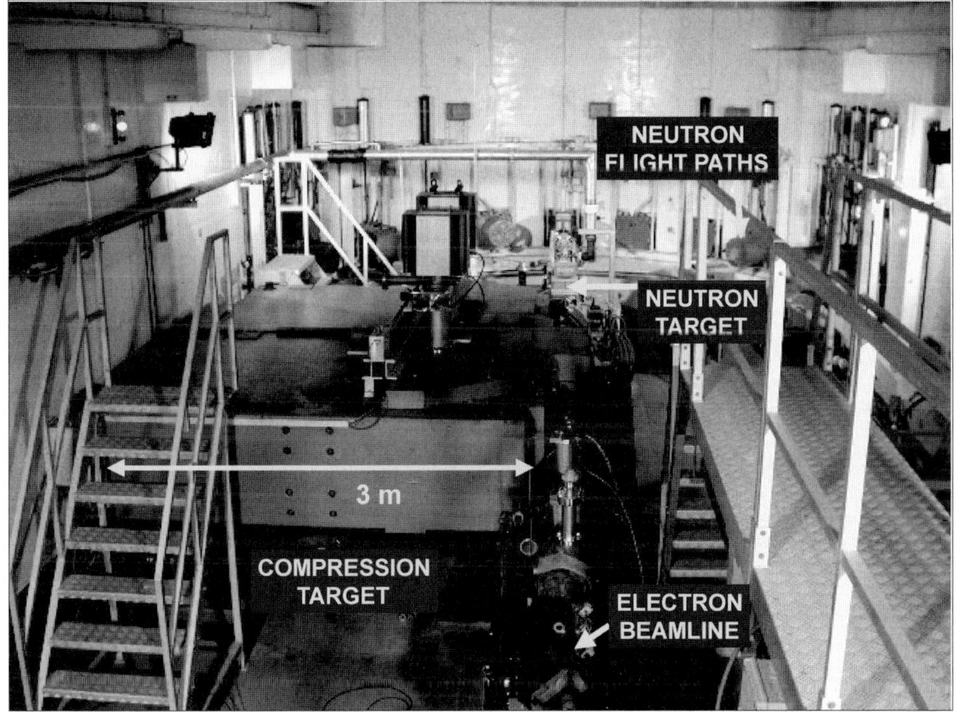

Figure 2.4. Location of the GELINA compression magnet in the target hall.

2.3   The GELINA neutron producing target

When the electron pulses, with energy uniformly varying from 70 to 140 MeV, hit the neutron producing target, bremsstrahlung is generated. The bremsstrahlung photons produce short bursts of neutrons mainly by (γ,n), (γ,xn) and (γ,F) reactions. Generally, the bremsstrahlung cross section is proportional to Z^2, which previously led to a conclusion in favor of using a heavy material as a target. Therefore, the rotary target currently in use consists of uranium-molybdenum (U-Mo) alloy with a 10%wt. of Mo, cooled by liquid mercury (Hg) and sealed in stainless steel. The U is depleted to 0.2%. Hg cooling is preferred over water cooling because neutron moderation needs to be avoided in the target itself. The whole target is 3-cm thick with a maximum diameter of 32.4 cm. Figure 2.5 shows the sketch of the present U-Mo rotary target. Hg flowing through the present target at a flow rate of 10 l/min is driven by an electromagnetic pump [Sal81], and is continuously removing the heat deposited in the target. Because the maximum electron beam power reaches 10 kW, the target rotates to improve the efficiency of power dissipation. The target rotates at a speed of 15 rpm.

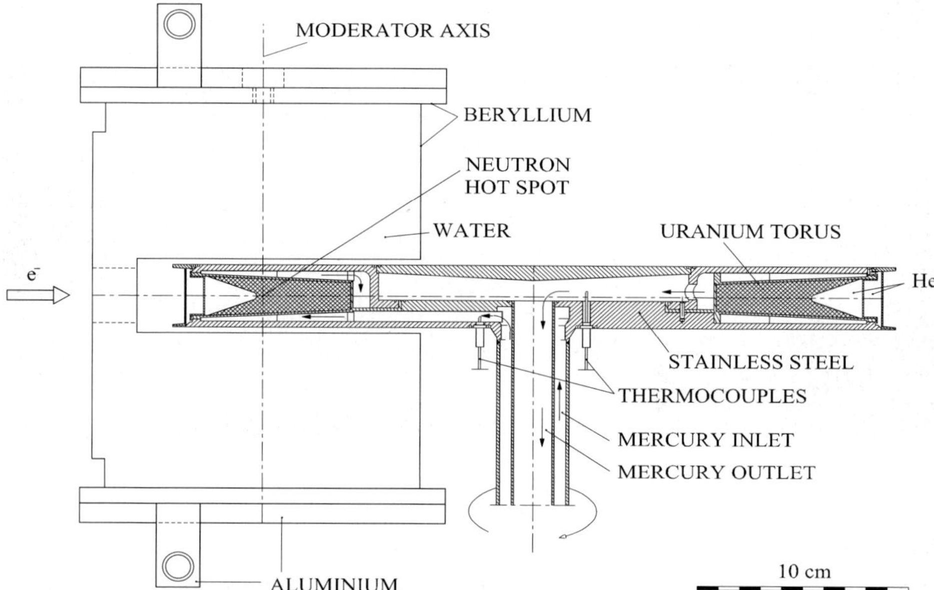

Figure 2.5. The present GELINA rotary target. Water tanks are placed above and below the target. Several thermocouples are used to monitor the temperature of the target.

The energy of the neutrons emitted by the target ranges from subthermal to about 20 MeV, with a peak at 1-2 MeV. When needed, fast neutrons produced by the target are subsequently moderated to lower energies by using two 4-cm thick beryllium (Be) tanks filled with water placed above and below the target (see Figure 2.6). In this manner, two neutron flux conditions are available: one optimized for energies below 100 keV by using the moderator tanks and one with fast neutrons coming directly from the uranium. Shadow bars can be placed between the source and the flight path to shield unwanted neutrons.

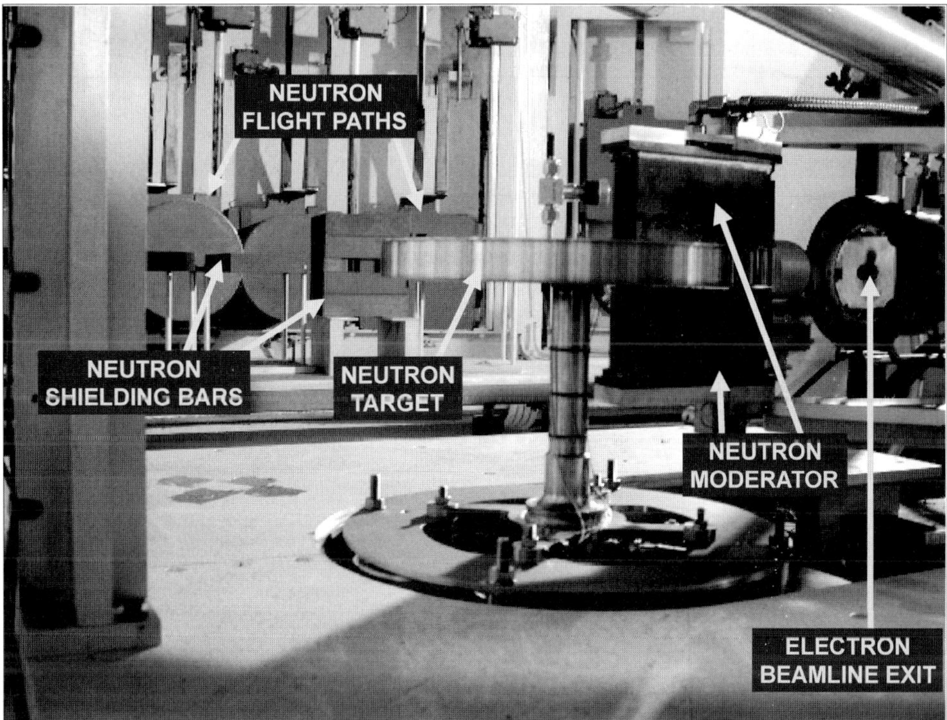

Figure 2.6. The GELINA rotary target in the target hall.

2.4 GELINA flight paths

In order to apply the TOF method at the GELINA in an efficient way, flight paths have been installed in a star-like configuration around the neutron producing target. Neutron measurements can be performed simultaneously at 18 flight paths with lengths ranging from 8 up to 400 m. The forward-peaked γ-flash favors flight path installation perpendicular to the electron beam axis. Figure 2.7 provides the top view on the location of the flight paths. Along the flight paths several measurement stations are installed at various distances. These experimental stations are fully equipped with a wide variety of sophisticated detectors, and data acquisition and analysis systems. Many types of neutron cross section measurements are possible: transmission experiments, and capture, fission, elastic and inelastic cross section experiments.

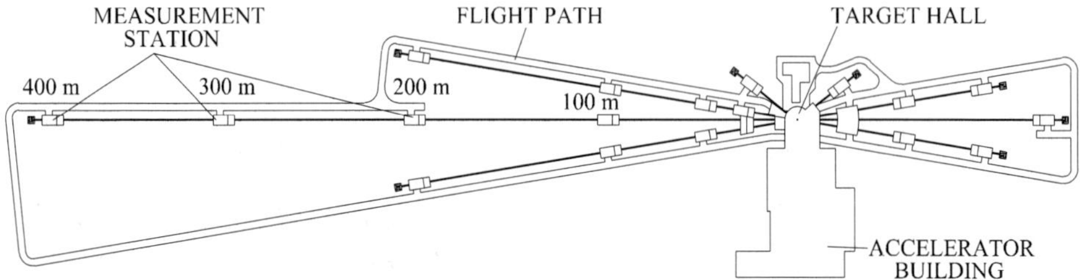

Figure 2.7. The top-view sketch of the GELINA site.

Chapter 3

Monte Carlo simulations for the GELINA rotary target

A design study was carried out with the aid of Monte Carlo techniques in order to come up with a final proposal for an improved neutron producing target. Since the GELINA facility uses electrons as initial particles to generate neutrons by gamma-induced nuclear reactions, coupled electron-photon-neutron transport simulations are required. The MCNP4C3 code [Brie00] is very well established for such transport calculations. This code is a general purpose, continuous energy, generalized-geometry, and time-dependent, particle-coupled Monte Carlo transport code.

Traditionally, in MCNP neutron-induced reactions were allowed to generate photons, but not vice versa. Since version 4C2 of the code, photons may also generate neutrons and it is this feature that makes it of direct use to study photoneutron targets such as used at GELINA. Several special MCNP features as flexible tallying and variance reduction techniques were applied to avoid approximations while maintaining calculation efficiency. Because photonuclear reactions are the source of the neutrons that are produced in the GELINA target, photonuclear libraries are required for the materials involved. Not all of these are included in the standard MCNP package so additional libraries were prepared from the IAEA evaluation.

In this chapter a comprehensive description is given of how the problem was approached. To verify the simulation approach, benchmark calculations have been performed for the existing target and for two neutron spectrum measurements at different flight paths. Also, RFs were calculated for several flight path angles, covering the energy range of interest. Since the RF is a very weak function of the neutron energy when expressed in terms of delay distance it was possible to apply broad energy bins when using Monte Carlo simulation techniques. The RFs were subsequently compared with earlier calculations using a code dedicated to the problem [Coc96]. Finally, the MCNP calculations were verified against RFs measured at two different resonance energies for the ^{56}Fe(n,γ) reaction.

3.1 GELINA neutron source modeling

The GELINA neutron source and the TOF facility were modeled with the MCNP4C3 Monte Carlo code [Brie00]. The use of this code involved the following requirements:
- to provide an appropriate geometrical model due to the complexity of the existing neutron target and of the flight paths (collimators, windows, filters, ...),
- to provide photonuclear data for all relevant nuclides involved in the problem,
- to implement existing variance reduction techniques to account for the poor statistics due to the long distances in the flight paths,
- to incorporate the RFs by introducing the delay distance concept.

3.1.1 MCNP4C3 model geometry

Due to the complexity of the present GELINA target, which is mainly dominated by the complicated shape of the Hg cooling channels, a geometry simplification was applied, based on the work performed by Coceva et al. [Coc96]. The introduction of the existing target is given in Chapter 2. Figure 3.1 shows the target geometry used for the MCNP4C3 simulations. The central cylindrical part of the target representing the cooling channels was modeled as a homogeneous mixture of Hg and Fe with a density of 11.02 g/cm^3. Hg annuli with a density of 13.3 g/cm^3, placed above and below the U-Mo part, substitute for the cooling channels of the real geometry. The same annular shape, but of different thickness, was used for the Fe constructional material with a density of 7.87 g/cm^3. Tanks of Be (1.848 g/cm^3) filled with water comprise the moderator located above and below the target.

Besides the physical model of the target, the collimators for the moderated neutron spectrum and for the unmoderated spectrum were included, using dry air, mylar windows and filters, as appropriate (see section 3.2).

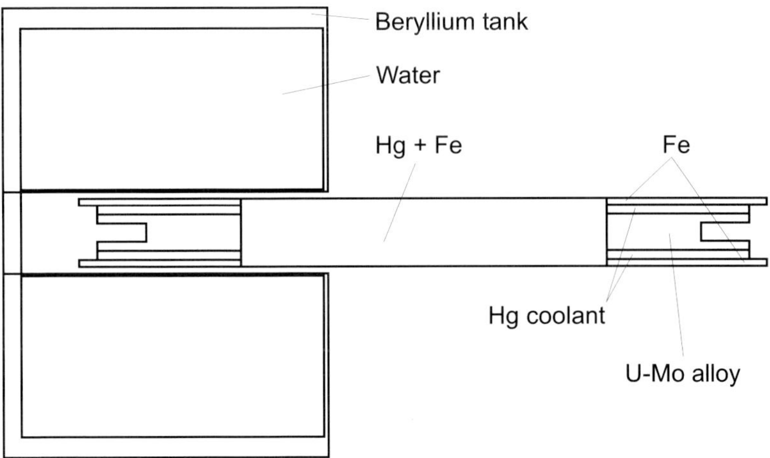

Figure 3.1. Side view of the MCNP rotary target model used in neutronics simulations.

3.1.2   *Photonuclear data*

3.1.2.1   Photoneutron production

The photonuclear cross sections as a function of photon energy are shown in Figure 3.2 [Vey73, Cal80b, Lep87, Rie84, Ost78, Kor79, Fro94]. In the photon energy range of interest (5-140 MeV) neutron production is dominated by the **Giant Dipole Resonance (GDR)** that manifests itself between 10 and 20 MeV [Vey73, Cal80a]. In the GDR region, neutrons are produced mainly through the (γ,n) and $(\gamma,2n)$ photoneutron reactions, and by the photofission reaction (γ,F) for ^{238}U. Above 20 MeV, in the so-called quasi-deuteron range, the total γ-induced reaction cross section is approximately constant and well below the values in the GDR range. In the case of the most important isotope, ^{238}U, this region is completely dominated by the (γ,F) process, i.e. about 90% of the cross section leads to fission. The neutron multiplicity ν in this range rises linearly from about 5 at 20 MeV to about 11 at 140 MeV [Ryc88].

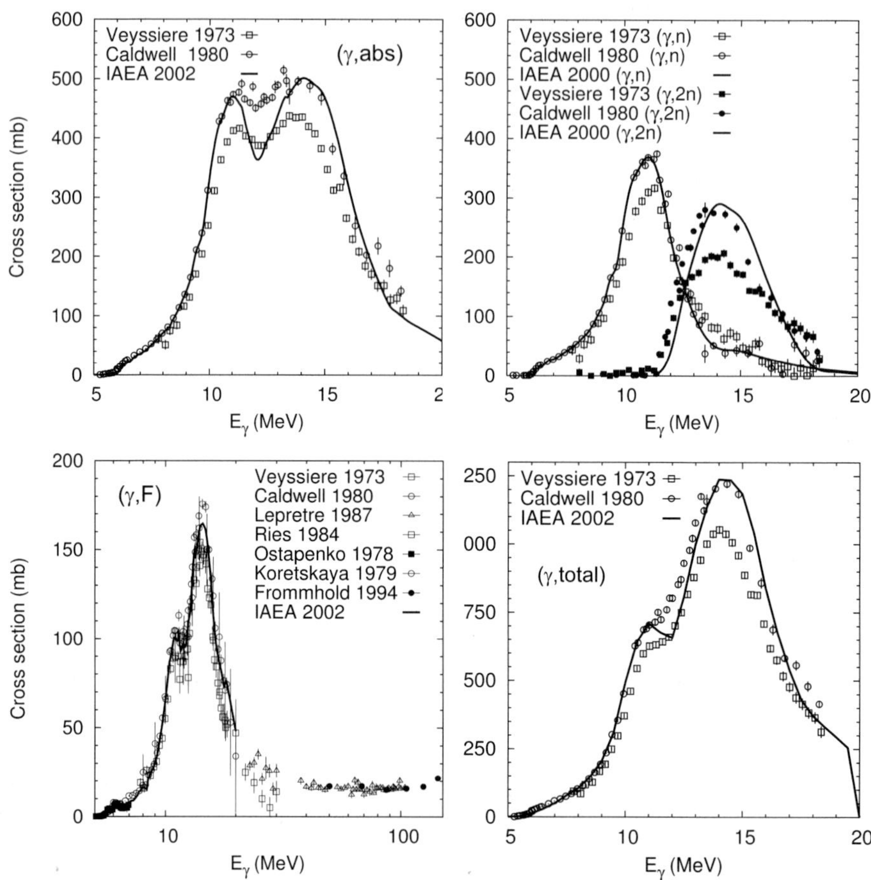

Figure 3.2. **Photonuclear cross sections of ^{238}U. Shown are the photoabsorption cross section (upper left), the (γ,1n) and (γ,2n) cross sections (upper right), the fission cross section (lower left) and the total neutron production cross section (lower right). The graph scales differ in some cases.**

Generally, the total neutron production cross section σ^{prod} is given by

$$\sigma^{prod} = \sigma(\gamma,n) + 2\sigma(\gamma,2n) + \cdots + \nu\sigma(\gamma,F) \qquad (3.1)$$

which takes into account the multiplicity of neutrons emitted in each process.

3.1.2.2   MCNP4C3 data libraries

A new photonuclear data library LA150u [Whi00] was used, which has been supplied with MCNP since version 4C2. It contains twelve photonuclear evaluations, originally in ENDF-6 format, which were processed into the ACE format needed for MCNP. From this data library, the photonuclear cross sections and neutron emission spectra of ^{16}O and ^{56}Fe were used. In case of ^9Be, ^{96}Mo and ^{238}U we obtained the photonuclear cross sections and neutron emission spectra from the IAEA data library containing 164 isotopes [IAE00]. This library consists of evaluations that are only available in ENDF-6 format. Therefore additional processing was made using NJOY99.90 in order to obtain the required ACE format libraries. Since neither of these libraries contains photonuclear data for Hg, the data for ^{197}Au were used instead. It was verified that the photonuclear cross sections for Hg and Au are very similar and therefore such a replacement is justified [Die88].

For the case of ^{238}U, the evaluation that was used is compared with the measured data in Figure 3.2. The evaluation that concerns the region below 20 MeV is the result of a statistical model calculation for which the parameters were determined from the data by Caldwell et al [Cal80a]. Therefore it follows closely the more recent (γ,1n) and (γ,2n) cross sections [Cal80b] and ignores the older set of Veyssiere et al. [Vey73]. The overall agreement with the data is rather good and it may be concluded that the cross sections are known to about 15-20% in relative terms. As shown in Figure 3.2, the library for U (curve IAEA 2002) assumes the cross section to be zero for energies above 20 MeV. This implies an underestimation of the total neutron yield. From the fission cross section and the above mentioned neutron multiplicity the neutron production cross section is expected to drop from 250 mb at 20 MeV to about 100 mb at 30 MeV, to be followed by a gradual rise to about 220 mb at 140 MeV. The spectrum of the emitted neutrons is also of relevance to the benchmarking in this work. The experimental data concerning the spectrum are very limited. The data file provides an evaporation spectrum for the (γ,1n) and (γ,2n) contributions. Each of these is characterized by its own nuclear temperature that depends on the incident energy. The fission neutron spectrum is characterized by a third energy-dependent temperature [Eva55].

3.1.3   *Variance reduction techniques*

Besides the features already quoted, an important advantage of the MCNP code is the flexibility it offers compared to codes dedicated to deal with one problem while maintaining state-of-the-art modeling of the physics processes. In particular, in MCNP it is quite straightforward to define geometry and choose materials that correspond as closely as desired to the experimental conditions. However, for an analog Monte Carlo model, which uses the natural probabilities that various events occur [Brie00] this would result in poor statistics, even after long running times on advanced computer systems. This is caused by very low efficiency of the MCNP calculations using the analog model due to experimental arrangements that are considered in this work. In order to obtain reliable results with a reasonable number of source particles variance reduction techniques were used to increase substantially the efficiency with respect to analog MCNP model calculations.

Two special variance reduction techniques were used. The first, called DXTRAN, was applied for the simulations of the moderated spectrum, while the simpler second technique, called point detector, was applied for the unmoderated spectrum. The point detector technique is in principle the same technique as DXTRAN, if the radius of the DXTRAN sphere is zero [Brie00]. For the case of a point detector, the contribution to the flux at a given point is determined for each neutron scattering. The "real" neutron continues normally, but the contribution is calculated for a pseudoneutron that scatters to the point. This pseudoparticle goes directly to the detector without any subsequent collision. Consequently, the weight of the contribution includes the differential cross section and the attenuation that may result between the point of interaction and the point detector. On the other hand, the DXTRAN technique is more useful than the point detector if near a physical detector, say inside a sphere of 10-cm radius surrounding it, further transport should be applied to the pseudoneutron. Like in the case of the point detector a pseudoneutron is generated at every neutron interaction point. The weight of such a particle directly corresponds to the probability that it will be scattered to the DXTRAN sphere and that it will not undergo any collision before reaching the detector.

In order to suppress a high variance of simulations, and to speed up the calculations, additional variance reduction techniques were applied. The detector diagnostic card limited the small contributions to the tally by playing the Russian roulette for all particles, of which the weight was lower than a defined fraction of the average tally contribution per history. Further, the detector contribution card saved computing time by reducing the number of the tally contributions from the unimportant geometry cells. Both these cards represent a very helpful complement to the point detector and DXTRAN variance reduction techniques [Brie00].

3.1.4   Other important aspects of the target modeling

For the water in the thermal energy range, the $S(\alpha,\beta)$ model was used [Brie00], which takes into account the effects of the chemical binding. This model is applied for neutron energies below 4 eV. In addition, for all MCNP4C3 calculations related to the neutronics problem, an energy cutoff of 5 MeV was applied in order to eliminate all electrons and photons below this energy. These particles cannot contribute to the neutron production, as the energy threshold for the photoneutron production for ^{238}U is about 5 MeV. This step helped us to significantly improve the efficiency of the simulations, since especially the treatment of low-energy electrons is very time consuming.

3.1.5   Implementation of the delay distance concept in MCNP4C3

Because the delay distance as such cannot be obtained by running the standard MCNP4C3, we applied the user-supplied subroutine called TALLYX. This subroutine allows the user to modify any tally, and also to create a user-defined quantity binning, so that the desired distribution can be obtained. The definition of the delay distance, which corresponds to the delay distance used in the TALLYX subroutine for our purpose, is introduced by Equation 1.4. The internal MCNP4C3 variables *vel*, *tme*, *xxx*, *yyy* and *zzz* were used to meet the requirements. Here *vel* stands for the velocity of the neutron on the detector, *tme* is the time at which neutron tallies, and *xxx*, *yyy*, *zzz* are the coordinates of the neutron that tallies.

3.2   Benchmarking of the MCNP model – absolute fluxes

In order to check the physical model of the system and the transport calculations, comparisons were made for two different neutron spectrum measurements: the first was done at an angle of 81° and a distance of 60 m with a moderated spectrum, the second at an angle of 90° and a distance of 200 m with an unmoderated spectrum. The DXTRAN technique was applied for the simulations of the moderated spectrum in order to integrate the possibly nonflat flux profile over the active area of the detector. The simpler technique, the point detector, was applied for the unmoderated spectrum. In the latter case the flux profile was verified experimentally to be flat within 3% so that the flux in the center corresponds to the flux averaged over the beam area.

All relevant experimental characteristics were taken into consideration and implemented in the MCNP4C3 simulation. These are the effective detector area, the accelerator current, the presence of air, and the shape, the location and the material composition of the collimators. The detector in the case of the moderated neutron spectrum measurement was simulated using cylindrical geometry with a diameter equivalent to the neutron beam diameter, and a thickness of 10^{-3} mm. The statistical accuracy of the results was kept below 5% per bin as prescribed in the MCNP manual [Brie00].

3.2.1   The moderated neutron flux

3.2.1.1   Experimental setup

The absolute neutron flux of the moderated spectrum was measured in the energy range from 25 meV to 200 keV by Borella et al. [Bor05]. The neutron spectrum was constructed from the measurements at the 60-m station (above 1 eV) and the 14-m station (below 1 eV). The flux was measured using the $^{10}B(n,\alpha)^{7}Li$ reaction with an ionization chamber with six homogeneously evaporated boron layers enriched to 94.1% in ^{10}B and a total ^{10}B thickness of 210 µg/cm^2 (see Figure 3.3). Neutron energies were determined using the TOF technique. Neutrons coming directly from the uranium target were eliminated using a 10-cm thick, 5-cm high lead bar that was placed in the flight path at 1.5 m from the target. A combination of Cu and borated polyethylene collimators ensures a neutron beam with a diameter of 74.5 mm at the location of the boron chamber. The diameter of the boron deposits is 84 mm so that the active layers intercept the full beam. During the measurement at the 60-m station the GELINA accelerator was operated at 100 Hz producing 10-ns electron pulses with an average current of 8.7 µA and a beam power of 0.87 kW. During the measurement at the 14-m station the GELINA accelerator was operated at 40 Hz with an average current of 3.5 µA. The overall accuracy of the spectrum is 11-15%; 5-11% is due to the flux measurement (statistics, amount of ^{10}B in the beam), 10% is due to the current normalization. The electron current is measured on the GELINA target.

It should be noted that background contributions were determined using black resonance filters. The attenuation of the filters between the black resonances was corrected for in the measured data. Attenuation of the neutron flux is further minimized by use of evacuated tubes, which are only interrupted at the experimental stations. Attenuation due to air (in total 6 m) and the mylar windows (1.4 mm in total) was not corrected for in the experimental data. This was taken care of in the MCNP4C3 model.

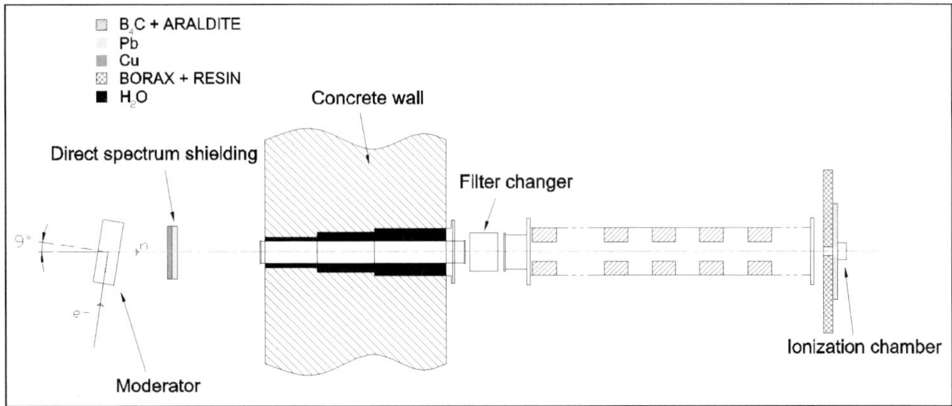

Figure 3.3. **The measurement configuration for the moderated flux measurement at 81° using the ionization chamber. Collimation systems were used to define the neutron beam.**

3.2.1.2 Results and discussion

Throughout the entire energy range, the agreement between the calculation and the measurement is very good in absolute values, as shown in Figure 3.4. The neutron flux is reproduced within 20%, which can be considered as a good agreement given the uncertainties stated above for the measurement and the photonuclear data (see section 3.1.2). The shape of the spectrum is very well described over the whole neutron energy range indicating the high accuracy of the moderation process modeling. For water in the thermal energy range, the S(α,β) and free gas models were compared. These models account for a thermal motion of atoms. In addition, the S(α,β) takes into account the effects of chemical bindings, and is applied for neutron energies below 4 eV. The comparison of the results reveals that the MCNP4C3 calculation of the thermal peak of the neutron spectrum below 0.5 eV is in better agreement with the measurement if the S(α,β) model is used. From 0.5 up to 4 eV the free gas model shows a slightly better agreement with the measured data.

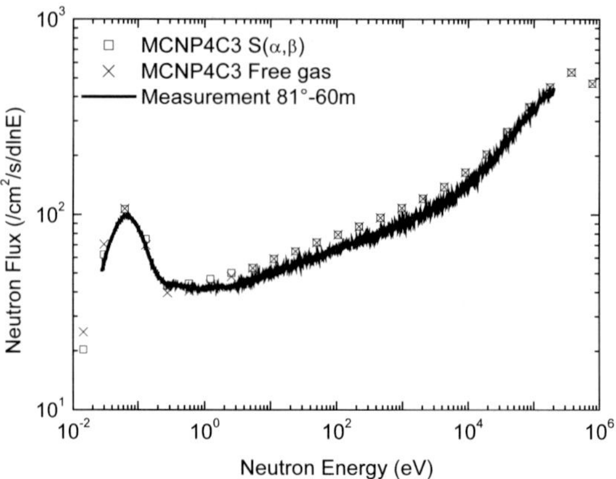

Figure 3.4. Absolute neutron flux per unit lethargy in the flight path 81° - 60 m of the moderated neutron spectrum.

3.2.2   *The unmoderated neutron flux*

3.2.2.1   Experimental setup

The absolute neutron flux of the unmoderated spectrum was measured in the energy range from 200 keV to 20 MeV by Mihailescu et al. [Mih04]. The measurement was performed using the ^{235}U(n,f) reaction with a fission chamber with eight homogeneously evaporated ^{235}U deposits (see Figure 3.5). Each deposit has 7-cm diameter and about 400 µg/cm^2 on aluminium backings with a thickness of 0.02 mm and a diameter of 8.45 cm. The ^{235}U is enriched to 99.826 +/- 0.008% and has a total areal density of 3.066 +/- 0.006 mg/cm^2. Following some pre-collimation the beam diameter of 61 mm was defined by a copper collimator located about 2 m in front of the fission chamber. Again, the beam was fully intercepted by the effective area of the fission chamber.

The moderated neutron spectrum was shielded using a 10-cm thick lead and a 10-cm thick copper shadow bar with a vertical opening of 3 cm and a horizontal opening of 12 cm. At the 100 m station two ^{10}B filters with a combined thickness of 1.23 g/cm^2 were placed to eliminate further low-energy neutrons, moreover a ^{nat}U disc of 36.137 g/cm^2 was used to block the intense prompt γ-flash. The filters ^{10}B and ^{nat}U decrease the flux by roughly a factor of two. These filters were included in the simulation. Neutron attenuation due to these filters, mylar windows and air was not corrected for in the measurement. During the measurement the accelerator was operated at 800 Hz with 1-ns electron pulses and a power of 7 kW. The flux measurement has an accuracy of about 4%, whereas the current normalization has an accuracy of 10%. The measurement corresponds to about 1000 hours of accelerator operation.

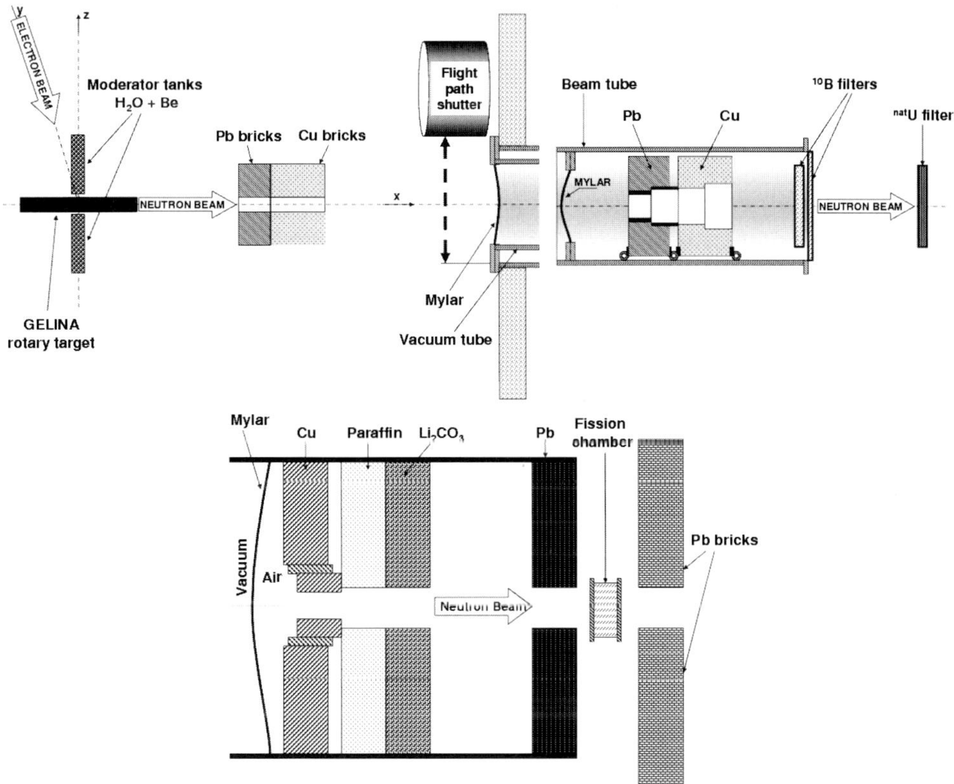

Figure 3.5. Neutron beam definition for the unmoderated flux measurement at 90° - 200 m. The rotary target with the first collimation system (upper left), the ^{nat}U and ^{10}B filters with another collimation system (upper right) and the beam defining collimators (bottom) are shown. These measurement configurations correspond to the target hall and the measurement stations at 100 m and 200 m.

3.2.2.2   Results and discussion

Figure 3.6 shows the comparison of the measurement with our MCNP4C3 simulation. The measurement covers the neutron energy range from 165 keV up to 20 MeV. In the energy region up to 5 MeV there is an agreement within 20%. Above this energy, an increasing deviation can be observed. This is probably caused by the (γ,xn) emission spectrum in the photonuclear data library for ^{238}U. The cause of the further apparent differences between the shape of the model calculation and the shape of the measured data was investigated. In order to verify the influence of the neutron inelastic scattering data of ^{238}U on the absolute flux simulation, the calculation was also done with the JENDL-3.3 library, which provides more accurate inelastic scattering data than the MCNP4C3 default (ENDF/B-VI). It can be seen in Figure 3.6 that the results are very similar with an observable difference in the region up to 400 keV only. This change does not account for the difference with the measured data. Also, replacing the simplified model of the GELINA target (Figure 3.1) by a model with the actual U-Mo conical shape (Figure 2.5) does not lead to a significant difference. Finally, it is expected that an overall increase by 20% of the temperature parameters that govern the neutron emission spectra would lead to an improved agreement with the shape of the measured data.

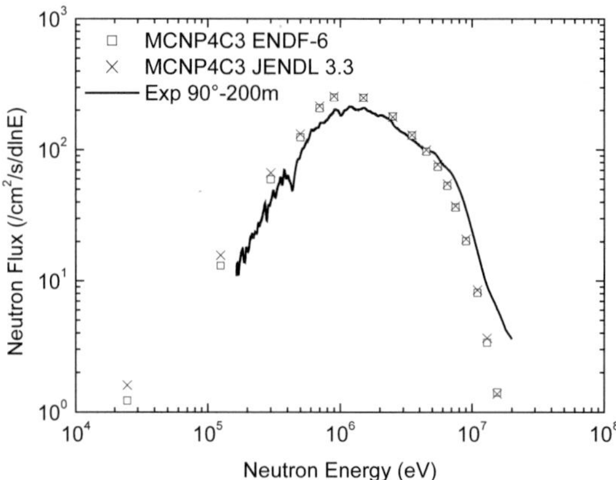

Figure 3.6.   Absolute neutron flux per unit lethargy in the flight path 90° - 200 m of the unmoderated neutron spectrum.

3.3 Benchmarking of the MCNP model – resolution functions

3.3.1 *Resolution functions of the moderated neutron spectrum*

A 1-ns electron pulse with a uniform electron energy distribution between 70 and 140 MeV was applied in order to obtain the neutron time response of the target. The neutrons arising directly from the target were shielded using a lead bar so that only neutrons coming from the moderator could contribute to the distribution. Further, a subdivision was done of the considered energy range from 1 eV up to 3 MeV into nine intervals, in which the RFs are only weak functions of the energy when expressed in terms of distances. In this way, the neutron response of the target on the electron pulse can be described. The RFs are presented in Figures 3.7 and 3.8, where the RF is given as a **Probability Density (PD)**, which is a function of the delay distance.

From the comparison it is apparent that the MCNP4C3 RF simulations are in all energy bins in very good agreement with those presently in use. The GELINA RFs were obtained in the past by Coceva et al. using a dedicated Monte Carlo code [Coc96]. The delay distance binning has the same structure in both cases. A logarithmic scale was used in order to see the slight differences for low PD values. The differences increase for the high neutron energies in the tail of the RF. The quality of the RF for the higher energies is of lower importance compared to those of lower energies, since the direct neutron spectrum is shielded. If the neutron cross section measurement requires high neutron energies, the fast neutrons are used instead of the moderated neutrons; this step is accomplished by shielding the moderator tanks. The results have relative errors less than 5%, which is a reasonable limit for use of the point detector in the MCNP4C3 code.

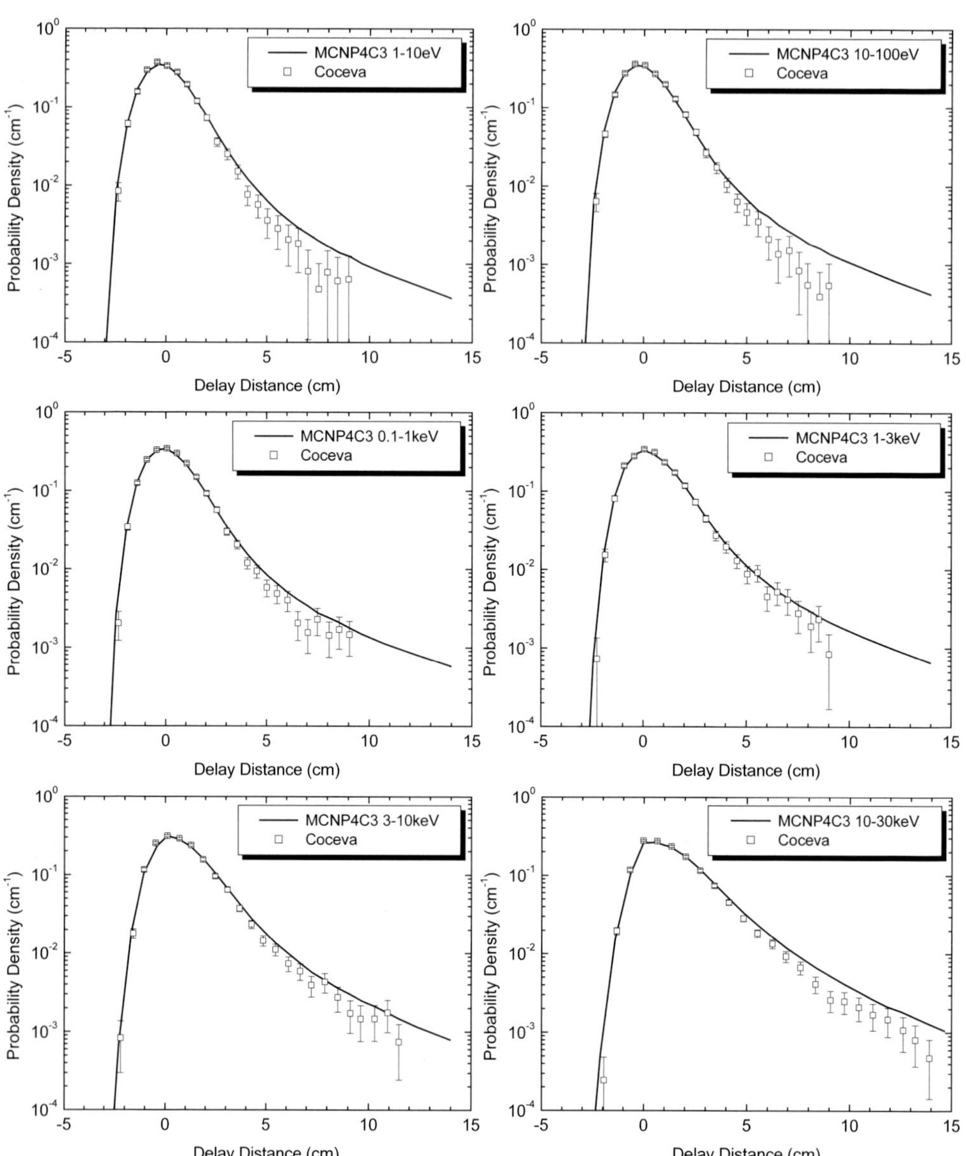

Figure 3.7. Resolution functions of the moderated neutron spectrum of the GELINA target in the flight path 81° and the energy range from 1 eV to 30 keV; relative errors of the MCNP results are below 5%.

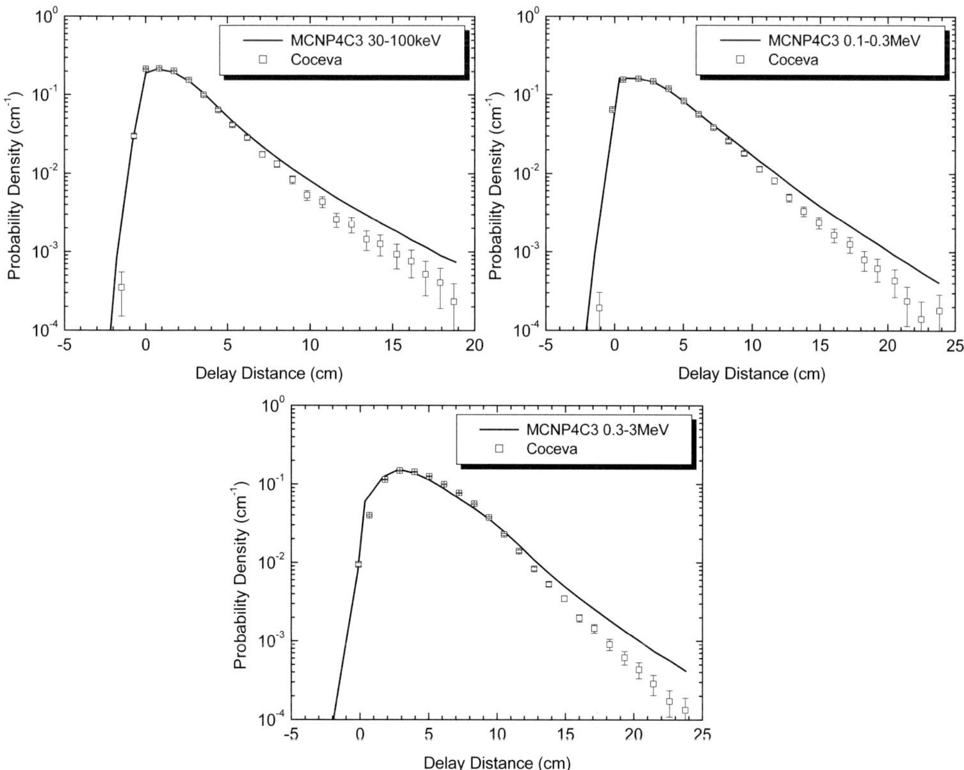

Figure 3.8. **Resolution functions of the moderated neutron spectrum of the GELINA target in the flight path 81° and energy range from 30 keV to 3 MeV; relative errors of the MCNP results are below 5%.**

As can be seen in the results, negative values of delay distances are obtained too. According to the specification of the delay distance by Equation 3.2, its definition is such that it also can be negative. This can be demonstrated by considering the distance l_i traveled by the neutron at the velocity v_i between collisions i-1 and i, where l_n is the distance until the point of the interaction in a hypothetical infinitely thin detector.

$$d = \sum_{i=1}^{n} \frac{v_n}{v_i} l_i - L \tag{3.2}$$

Explicitly,

$$l_n = L - \frac{T}{2} + \Delta l_n \tag{3.3}$$

where T is the thickness of the moderator and Δl_n is the distance traveled by neutron in the moderator since the last collision in the moderator. Thus,

$$d = -\frac{T}{2} + \Delta l_n + \sum_{i=1}^{n-1} \frac{v_n}{v_i} l_i \tag{3.4}$$

It is evident that d can be negative when $v_n \ll v_i$ for the largest part of the distance traveled in the moderator and $\Delta l_n < T/2$. It is also clear that the RF distribution must shift towards increasingly positive values when the moderated neutron energies are closer to the initial neutron energies.

3.3.2   Resolution function of the direct neutron spectrum

In these simulations the same conditions for the electron beam have been applied as in section 3.3.1. However, the shielding bar was placed such that it blocked all neutrons coming from the moderator. Therefore only neutrons coming from the rotary target could reach the detector. The detector was placed at an angle of 90°. The energy range from 3 keV to 20 MeV was considered. The results are depicted in Figures 3.9 and 3.10. Again, the RFs are expressed in terms of the delay distance. The results represent the behavior of the neutrons of given energy bins during the slowing-down process in the rotary target. The relative errors are shown in the results. As the energy of the neutrons increases, the peak of the neutrons scattered by uranium located around d = 0 is more apparent, especially in the energy bins up to 1 MeV. The additional tail of the RF is created by the neutrons, which underwent elastic scattering collisions in the moderator, and subsequently were scattered back to uranium before leaving the target. At high energies we can clearly observe the RF distributions, which correspond to the 1-ns electron pulse and the varying birth locations of the neutrons.

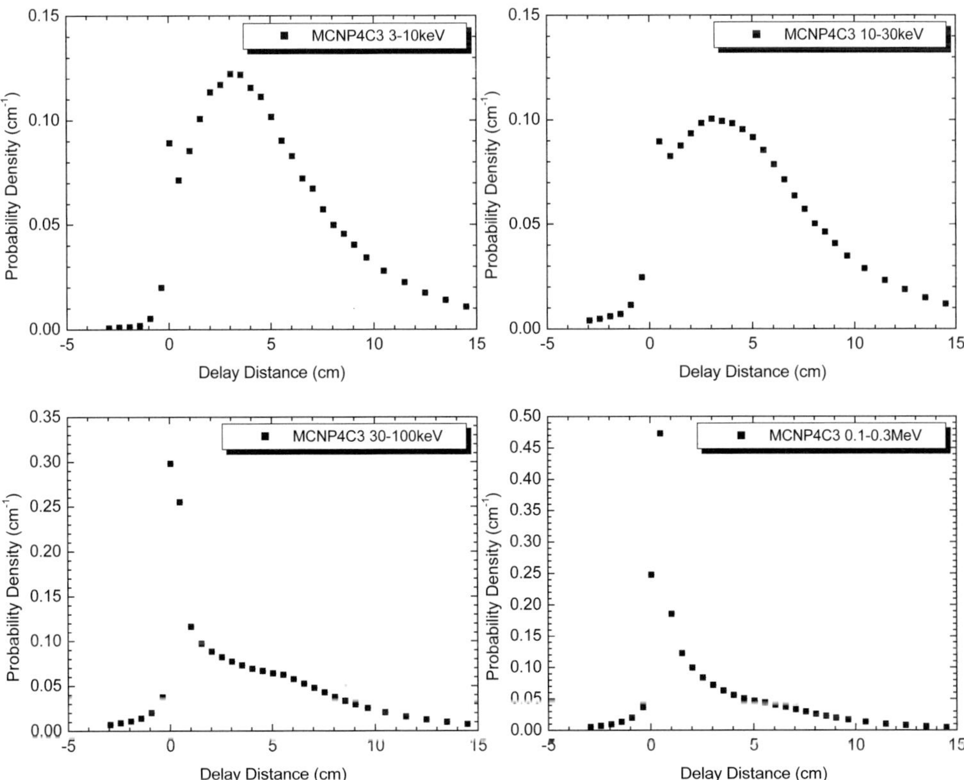

Figure 3.9. Resolution functions of the direct neutron spectrum of the GELINA target in the flight path 90° and energy range from 3 to 300 keV; relative errors are shown, but are smaller than the points.

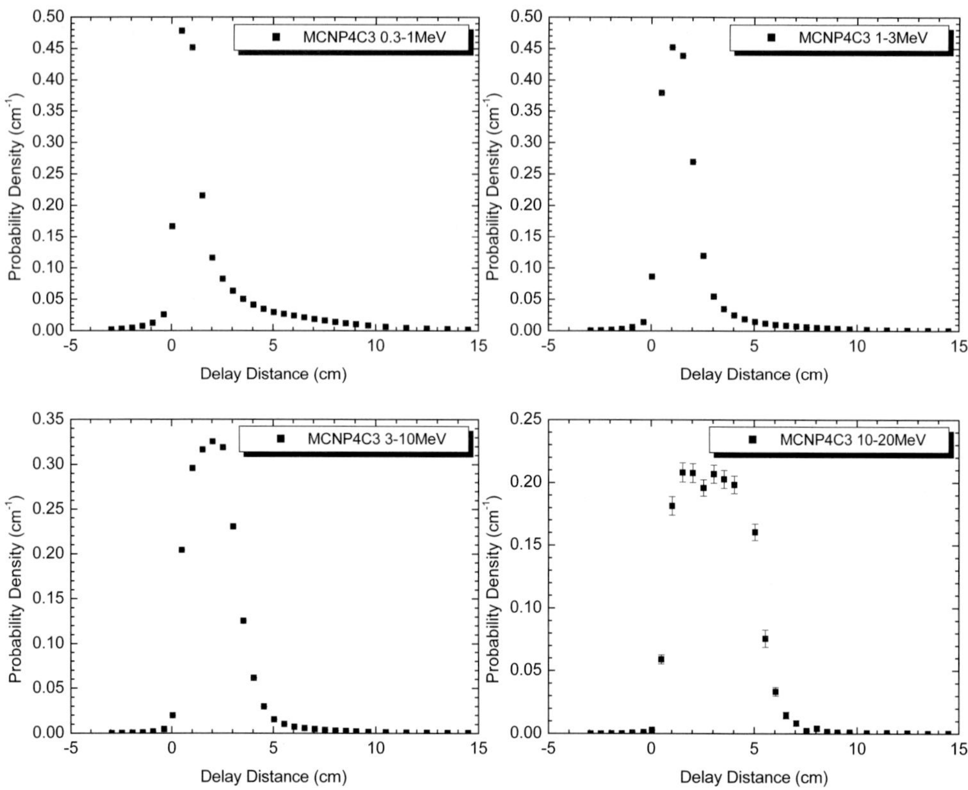

Figure 3.10. Resolution functions of the direct neutron spectrum of the GELINA target in the flight path 90° and energy range from 0.3 to 20 MeV; relative errors are shown, but in some cases are smaller than the points.

3.3.3   *Comparison of the resolution functions for different angles*

As a last step of the investigation of the RFs of the existing rotary target, a comparison of the RFs has been carried out for different flight paths, namely for 54°, 72°, 81°, and 90°, with respect to the electron beam. It needs to be emphasized that only the moderated neutron spectrum was used in these simulations. The results are shown in Figure 3.11 for two different energy bins. While the shape of the tail of the RF curves is not significantly changed in the range of interest, the peak of the curves is broadened significantly with decreasing angle. This phenomenon is evident especially for low neutron energies as there is a significant difference between the neutron energy at birth and the energy after the last collision in the moderator before leaving the moderator towards the detector. To explain this angle dependency, it has to be realized that a geometrical effect, originating from the shape of the moderator, plays a role for angles different from 90° (see Figure 3.12). This geometrical effect is negligible for 90° if the size of the moderator is much smaller than the source-detector distance. In such a case all neutrons of the same energy travel the same distance to the detector after leaving the moderator. However, if the detector is located at a different angle than 90°, the distance traveled by a neutron of given energy depends on the location of the neutron exit point from the moderator.

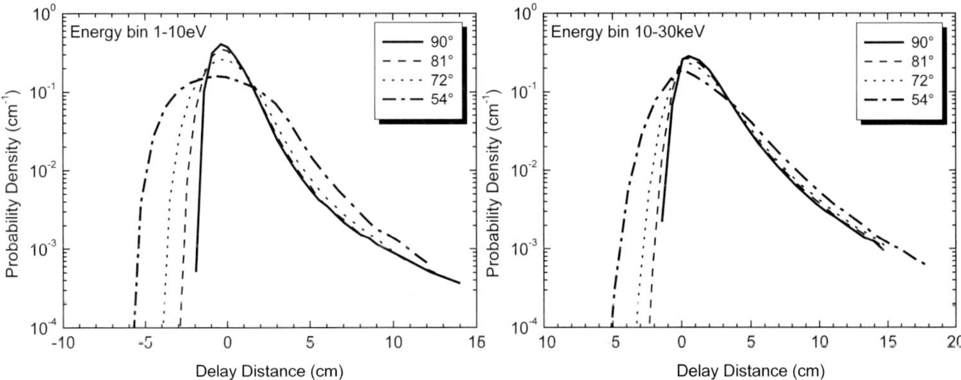

Figure 3.11. Resolution functions of the moderated neutron spectrum of the GELINA target for angles 90°, 81°, 72°, and 54°; relative errors are below 5%

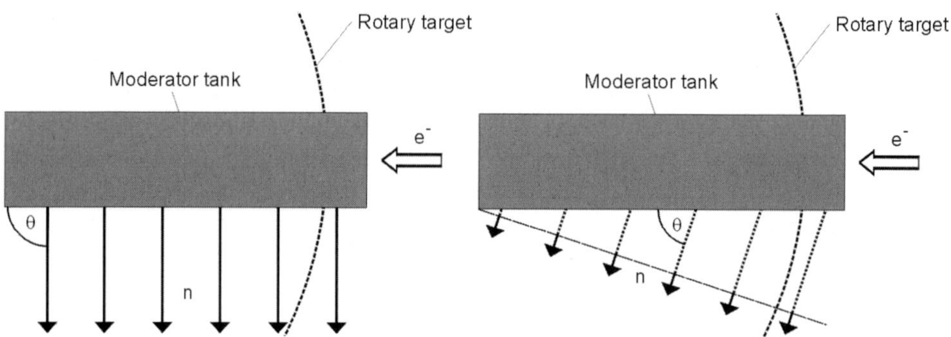

Figure 3.12. Top-view detail of the GELINA target-moderator assembly. If θ = 90° (left), no geometrical effect appears. If θ ≠ 90° (right, θ = 81°), neutrons have to travel different distance to detector after leaving the moderator, depending on the exit point of a neutron from the moderator.

Generally, the shape of the RFs is given by the combination of three interrelated effects: the geometrical effect, the scattering in material, and the spread of the last-collision points of the neutrons before leaving towards the detector. Figure 3.13 depicts the dependency of the FWHM values of the RFs on the flight path angle and the neutron energy. It emerges from the results that the FWHM increases with increasing $\cos\theta$ for both energy bins shown. This behavior is the result of the increasing importance of the geometrical effect with decreasing angle. As can be also seen in Figure 3.13, the angle dependency is stronger for low energies. This fact is caused by the spread of the exit points of neutrons from the moderator along the moderator surface. Naturally, low-energy neutrons undergo on average more collisions than neutrons with higher energies, and therefore there is a higher chance that these neutrons reach the periphery of the moderator. Consequently, the increase in FWHM is smaller for higher energies.

Figure 3.13 also shows how the FWHM values change with the neutron energy. To understand these results one should be aware of the fact that the elastic cross section for H is constant in the range of 1 eV – 5 keV, and has a value of about 20 barn (1 barn = 10^{-24} cm²). This is clearly exhibited in the result for 90°, in which the FWHM is roughly constant below 1 keV. Again, the geometrical effect is negligible in this case. The cross section for O equals to 3.8 barn in this energy range, and therefore can be neglected as well. At higher energies, the FWHM values increase as a result of decreasing cross section for H (4.3 barn at 1 MeV), which causes larger **mean-free-path** (mfp) values. The mfp between two collisions in H is 0.75 cm for 20 barn and 3.47 cm for 4.3 barn, respectively (6.7 x

10^{22} cm^{-3} of H nuclei). However, at high energies the O nuclei cannot be neglected anymore as they significantly contribute to the neutron mfp (mfp for O is 3.73 cm at 1 MeV). Consequently, energetic neutrons travel a longer distance between the two collisions. This increase of the mfp allows the neutrons to reach the detector by crossing a larger distance without collision in the moderator. This causes additional broadening of the RFs, having as a consequence higher FWHM values.

In the case of 54°, the geometrical effect is strong, and causes high FWHM values even for low neutron energies. This makes the results very different compared to the case of 90°. From about 100 keV, however, both curves behave in similar way. This energy region is dominated by few collisions of relatively long mfps.

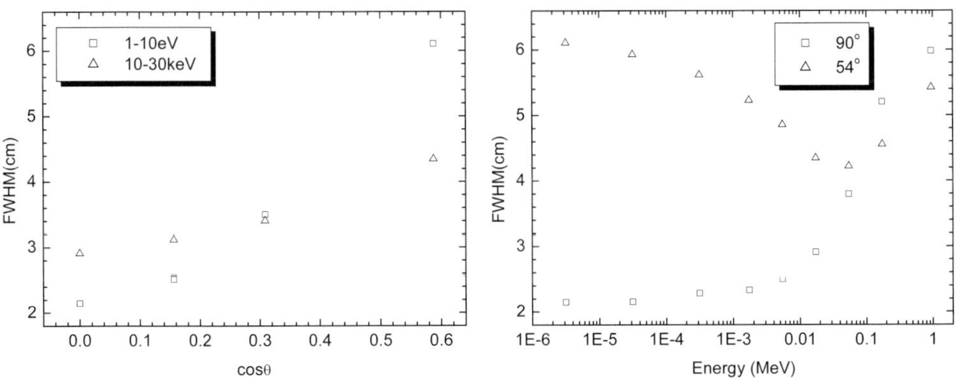

Figure 3.13. FWHM values of the RFs of moderated neutron spectrum, shown as a function of the flight path angle (left) and the energy (right).

3.3.4   *Comparison of the calculated and measured resolution functions*

The quality of the simulated RFs can be demonstrated for nuclides with very small intrinsic resonance widths in the energy region where the RF of the target outweighs the Doppler broadening. In such a case only the experimental broadening is important. The experimental broadening consists of the target broadening and the broadening of the measurement equipment. For a capture measurement the latter is negligible, thus only the target broadening is of the importance.

This situation occurs in a measurement that was carried out to obtain the shape of two resonances of ^{56}Fe using the (n,γ) reaction, namely at energies of 22.7 keV and 46 keV (see Figure 3.14). This capture measurement was performed with two C_6D_6 scintillation detectors having a fast time response so that the contribution of the measurement equipment to the resolution is negligible. The rest of the experimental setup was identical to the setup described in section 3.2.1. Two resonance fits of the isotope ^{56}Fe are shown in Figure 3.14. The FWHM values represent the partial contributions for the given resonance, namely the $FWHM_D$ for the Doppler effect and the $FWHM_R$ for the RF. The value Γ_{tot} is the FWHM of the natural resonance. One observes that the combined width $FWHM_{D+R}$ is completely dominated by the RF of the target. The fit (FIT curves) of the measured data (EXP) was performed by using the resonance shape analysis code REFIT_IRMM [Mox89]. This code is a general resonance fitting program, which takes into account all important physical phenomena. The code uses Coceva's RFs obtained in the past by Monte Carlo simulations, using the dedicated code [Coc96]. These RFs are shown in Figure 3.14 as provided by the REFIT_IRMM code. As concluded in section 3.3.1 there is only a small difference between the MCNP4C3 RF calculations and the Coceva's results, which appears mainly in the tail of the RFs below the **Full With** at one **HU**ndredth of the **M**aximum (FWHUM). From Figure 3.14 it can be deduced that these minor differences, appearing for long delay distances are not important. The energy axis (E_n in eV) in Figure 3.14 was derived from a time of flight t (in µs), using $E_n=(72.3L/t)^2$ where L (in m) is the flight path length. Naturally, the large delay distance values of the simulated RFs occur in the energy range before the resonance area where the resonance is flattened, and therefore small differences have no significant importance for the resonance shape obtained by fitting. In order to illustrate the low importance of these differences, Figure 3.14 shows the energy range defined by the FWHUM values of the MCNP4C3 RFs. It is obvious from this energy range and the location of the RF lines that no RF differences are important below the FWHUM.

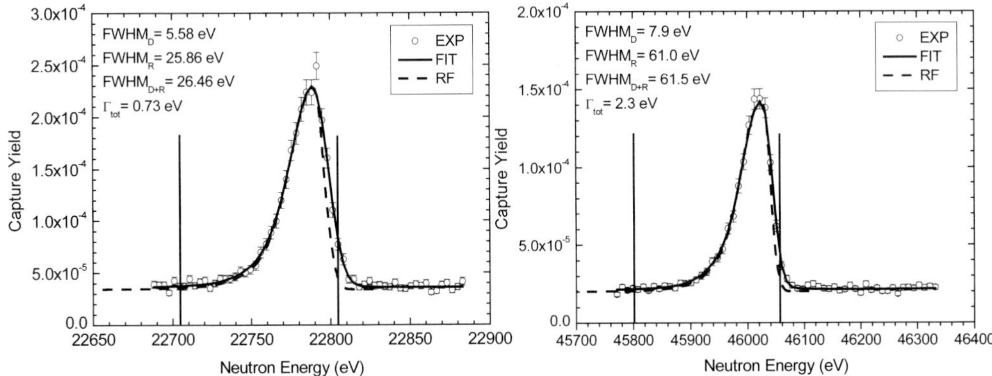

Figure 3.14. Resonance fits for the isotope [56]Fe obtained by the REFIT_IRMM code. The resonances at energies of 22.7 keV (left) and 46 keV (right) are shown. Vertical lines represent the energy range delimited by the FWHUM values of the MCNP4C3 RFs.

3.4   Conclusions

In order to model accurately the GELINA white spectrum neutron source with the MCNP4C3 code, required photonuclear data libraries were used. The neutron flux measurements at different flight paths were compared with the MCNP4C3 simulations. The results show a good agreement within the uncertainties of the measurements and the photonuclear data. Further, resolution function simulations were carried out for the GELINA rotary target, and compared with earlier calculations. Again, the results show a very good agreement within uncertainties, and demonstrate a high accuracy of the resolution functions presently used at IRMM. Therefore it can be concluded that the MCNP4C3 code together with the new photonuclear data libraries are a very suitable tool to design a new GELINA target.

Chapter 4

Influence of the target properties on the neutron production and the resolution functions

As mentioned in chapter 1 of this thesis, there has been a continuous effort at GELINA to improve the neutron production and the RFs over a broad energy range. In order to tackle such a problem it is essential to understand its physical background. Therefore, the aspects affecting the neutron production as well as the shape of the RFs are discussed in this chapter. An explanation is given of the approach chosen to investigate the optimal shape of the target from a neutron-yield point of view. Then the influence of the target size on the RFs is discussed in detail.

4.1   Optimization of the target size – neutron-yield point of view

At GELINA the neutrons are produced by high-energy electrons bombarding the rotary target. This process occurs via bremsstrahlung radiation and subsequent photonuclear reactions. Due to the fact that, at high energies, electrons and photons interact with matter mainly through bremsstrahlung and pair production, an electromagnetic cascade is induced in the target material. Intuitively, the shape and size of the cascade should be chosen as an inspiration for the choice of a new target shape to optimize the neutron production.

4.1.1   *Particle energy loss in medium*

When passing through matter, electrons lose energy to atomic electrons via electron-electron and electron-nucleon interactions. Two types of energy loss processes occur in electron collisions: radiative loss (bremsstrahlung) and ionization loss. In the high-energy range, electrons lose most of their energy via radiation, and ionization losses are negligible. Radiation losses are caused by the deceleration of electrons in the stopping material.

Generally, the rate at which a particle loses energy per unit path length is known as the stopping power of the medium. The Bethe-Bloch formula [Lil01] represents the mean ionization energy loss of a charged particle:

$$\left(\frac{dE}{dx}\right)_{ion} = -\left(\frac{q_c^2}{4\pi\varepsilon_0}\right)^2 \frac{4\pi Z\rho N_A}{Am_e v^2}\left[\ln\left(\frac{2m_e v^2}{I}\right) - \ln\left(1-\beta^2\right) - \beta^2\right] \qquad (4.1)$$

where $v = \beta c$ is the particle velocity, and c is the speed of light, q_c is the particle charge, ε_0 is the vacuum permittivity, m_e is the electron mass, N_A is Avogadro's number, Z, A, and ρ are the atomic number, mass number and density of the medium, respectively, and I is the mean ionization energy of the medium atom. I can be determined empirically as $I = 11Z$ (eV) [Lil01]. It should be emphasized that the quantum-mechanical derivation of Equation 4.1 includes relativistic effects, which take place for high particle velocities. However, for particles with velocities much lower than c the terms β^2 may be neglected in Equation 4.1.

On the other hand, energy loss of the electron due to bremsstrahlung radiation is well approximated by the Bethe-Heitler equation [Per00]

$$\left(\frac{dE}{dx}\right)_{rad} = -E\frac{4N_A Z^2 r_e^2}{137A}\ln\left(\frac{183}{\sqrt[3]{Z}}\right) \qquad (4.2)$$

where E is the initial energy of the electron, x the areal density (g/cm²), and r_e is the classical electron radius.

The development of the electromagnetic cascade reaches a maximum when the particles reach the critical energy E_C. At this point, the radiation and ionization losses are equal. E_C can be expressed using Rossi's definition for solid materials [Ros52]

$$E_C = \frac{610\,MeV}{Z+1.24} \qquad (4.3)$$

For electron energies above E_C, the electron energy loss is dominated by bremsstrahlung, whereas below it is dominated by ionization and excitation.

4.1.2 Electromagnetic cascade

When a high-energy electron initiates an electromagnetic cascade in a material, many electrons and photons with lower energy are generated via pair production and bremsstrahlung. The number of particles increases steeply as the energy of the initial electron is divided among the secondary particles. As a result, the cascade broadens as it develops.

As mentioned earlier, the shape and size of the cascade could give us a lead what the new target should look like. The range of an electron cannot be defined so well as the range of heavy ions due to the production of the secondary particles. An electron can lose a large fraction of energy in a single electron-electron collision. Moreover, the electron which penetrates further into the medium may not be the incident one. Therefore, quantities representing the mean behavior of the electromagnetic cascade are of the interest. Generally, three quantities are sufficient to describe the electromagnetic cascade. They will be used further for the scaling of the target dimensions.

For the scaling of the longitudinal cascade development, which is governed by the high-energy part of the cascade, the material radiation length X_0 can be introduced. The expression

$$\left(\frac{dE}{dx}\right)_{rad} = -\frac{E}{X_0} \tag{4.4}$$

defines X_0 as a mean distance over which the high-energy electron loses all but $1/e$ of its energy by bremsstrahlung radiation [Bar96]. The radiation length X_0 is energy independent and can be directly calculated from Equation 4.5 (X_0 in cm) by Dahl [Bar96], in which the functional dependence on Z is pointed out.

$$X_0 = \frac{716.4 A}{Z(Z+1)\rho \ln\left(287/\sqrt{Z}\right)} \tag{4.5}$$

The Molière radius R_M describes the lateral spread of the shower. It is defined as [Nel66]

$$R_M = \frac{E_S}{E_C} X_0 \tag{4.6}$$

where E_S = 21 MeV is a constant of multiple scattering theory [Bar96]. Measurements of the lateral distribution of electromagnetic cascades have been carried out in the past [Nel66, Bat70]. On the average, 10% of the cascade energy lies outside the cylinder with radius R_M. Within a radius of $3.5R_M$ about 99% of energy is contained. The cascade starts from a narrow volume and broadens as it develops, and is symmetrical along the axis defined by the direction of the initial particle of the cascade [Gri89].

4.1.3   Target features affecting neutron production

Inspired by the shape of the electron-photon cascade, a cylindrical geometry has been chosen for initial investigation of the new target design. In this way the cascade of high-energy photons capable of producing neutrons can be fully embraced by the target without extensive use of material. A 4π neutron escape has been calculated for the existing GELINA rotary target to be about 5.6% n/e, using a pencil (infinitely narrow) electron beam. Consequently, a decision has been made to investigate the amount of material needed for a single-piece target to produce the same amount of neutrons.

In Figure 4.1 a comparison is shown of the 4π neutron production, escape and escape probability for U-Mo cylinders of various sizes, calculated by MCNP. U-Mo alloy with a 10%wt. of Mo has been used in these initial tests as the rotary target currently in use also consists of this alloy (see chapter 2). In the MCNP simulation, a pencil electron beam parallel to the axis of cylinder, with a uniform energy distribution between 70 and 140 MeV bombarded the target at the centre of the base. In contrast, the GELINA facility operates with an electron beam with a diameter of 1 cm. However, it has been verified that a narrow beam can be used instead without compromising the results. Naturally, the neutron escape has been calculated by subtracting the neutron absorption from the neutron production.

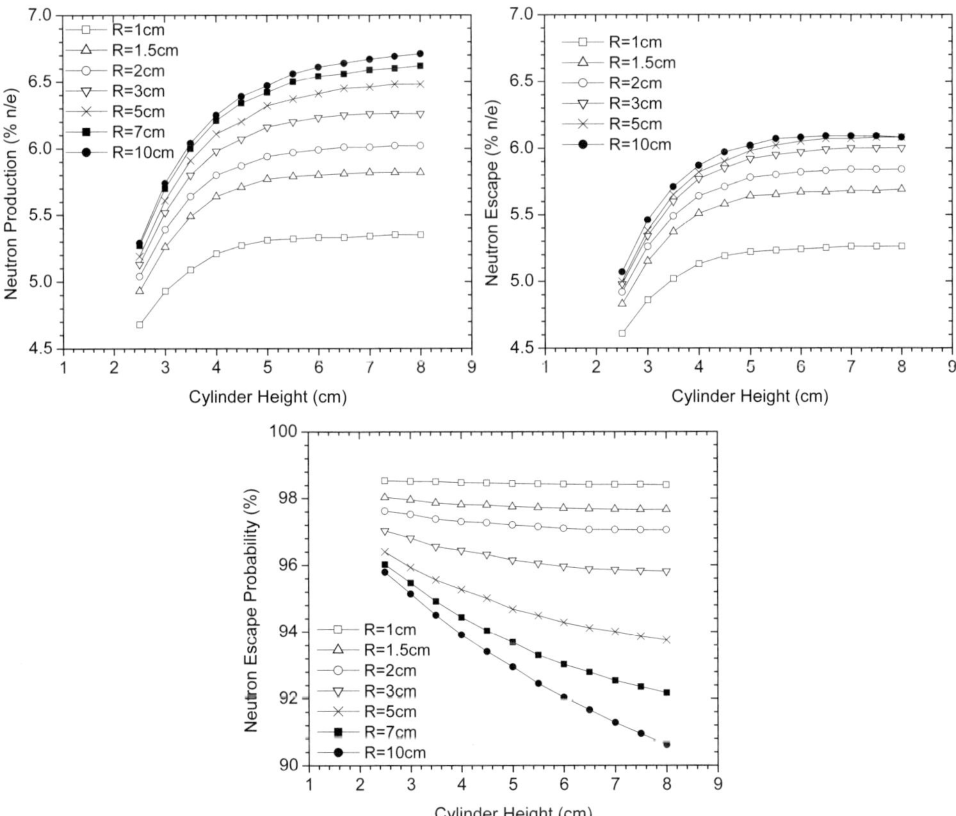

Figure 4.1. **Neutron production, neutron escape and neutron escape probability of the U-Mo cylinder for various heights and radii; relative errors < 5%.**

Above a certain size of the target, extra material behaves as a pure absorber, which does not significantly improve the neutron production. Therefore, at some point, the neutron production will saturate as the longitudinal and radial directions will increase. As can be seen in Figure 4.1, a tendency to saturate is observed in the longitudinal direction above H = 6 cm. A similar saturation effect is also apparent with increasing radius of the cylinder, although one would need to increase further the radius of the cylinder to reach the full-saturation point. However, the neutron-production curve for R = 10 cm, if compared with the one for R = 7 cm, indicates that this saturation point lies close to R = 10 cm. It can be concluded that the contribution to the neutron production per cm decreases with increasing R. This is the result of decreasing number of high-energy photons capable to generate other neutrons at the periphery of the electromagnetic cascade.

From Figure 4.1 it is apparent that neutron escape from the cylindrical geometry can be larger than from the GELINA target (5.6% n/e). The MCNP calculations reveal that the 4π neutron escape of the present target can be fully maintained by a cylinder with R = 1.5 cm, and H = 5 cm. This geometry produces slightly more neutrons than the rotary target.

It is observed that the neutron-escape curve corresponding to R = 5 cm is already very close to the curve R = 3 cm, thus, a similar saturation takes place for neutron escape as described above for the production of neutrons. This saturation effect is highlighted by the curve R = 10 cm, which is almost identical with the curve R = 5 cm, starting from H = 7 cm. This curve descends slightly from H = 7.5 cm (6.1% n/e), and reaches a value of 6% n/e at H = 15 cm (not shown in Figure 4.1). Clearly, the additional material absorbs more neutrons than it creates. Figure 4.1 also introduces the escape probability of various cylinders. This probability was calculated as a ratio of the neutron escape and production in %. For small R-values, the neutrons leave the geometry mostly in radial direction and therefore escape probability curves are flat. On the other hand, for high R-values, most neutrons leak in the axial direction. As a result, the escape probability decreases rapidly with increasing H. Finally, it can be concluded that the neutron escape of the single-piece target may reach the value of about 6.1% n/e when using U-Mo alloy, with the production reaching more than 6.5% n/e.

4.2   Optimization of the target size – resolution-function point of view

The size and shape of the actual rotary target and of the moderator are the major sources of RF broadening and consequently of neutron energy uncertainties. Therefore, to reveal clearly the relevant features of the target-moderator assembly dedicated MCNP simulations have been carried out. This investigation covers the entire neutron energy spectrum.

4.2.1   *Target features affecting resolution functions – neutron high-energy range*

In order to assess the influence of the properties and quantity of the target material on the RFs, with minimal influence of the target shape, a spherical geometry has been chosen for the simulations. An electron source producing a 1-ns pencil pulse with a uniform electron energy distribution between 70 and

140 MeV was placed at the centre of a U-Mo sphere to obtain the neutron time response of the geometry. For the initial simulation, the radius of the U-Mo sphere was set to 2 cm, with a point detector located 60 m from the sphere's centre, and at 90° with respect to the electron-beam direction. Subsequently, the radius of the sphere was changed to 4, 6, and 8 cm.

The results shown in Figures 4.2 and 4.3 are presented not only in a linear scale but also in a logarithmic scale so that the differences in the tail of the RFs are clearly visible. Accordingly, we used both scales also in other figures containing the RF distributions. In Figures 4.2 and 4.3, the influence of the amount of material in the target is manifest. While in the case of R = 2 cm and an energy bin of 100-300 keV a sharp RF with FWHM equal to about 1 cm can be observed, in cases having a larger amount of material the RF is drastically broadened. The tail broadening is more apparent with decreasing neutron energy and is caused by the neutron scattering process. In some cases this process results in the contraction of the flight path distance, which can be observed as the negative delay distance values. The results shown in Figures 4.2 and 4.3 indicate the necessity to reduce the target volume, if improved RFs are to be achieved for high neutron energies. Note that the RFs are normalized to unit integral over delay distance, and carry no information about the neutron yield in a given energy bin.

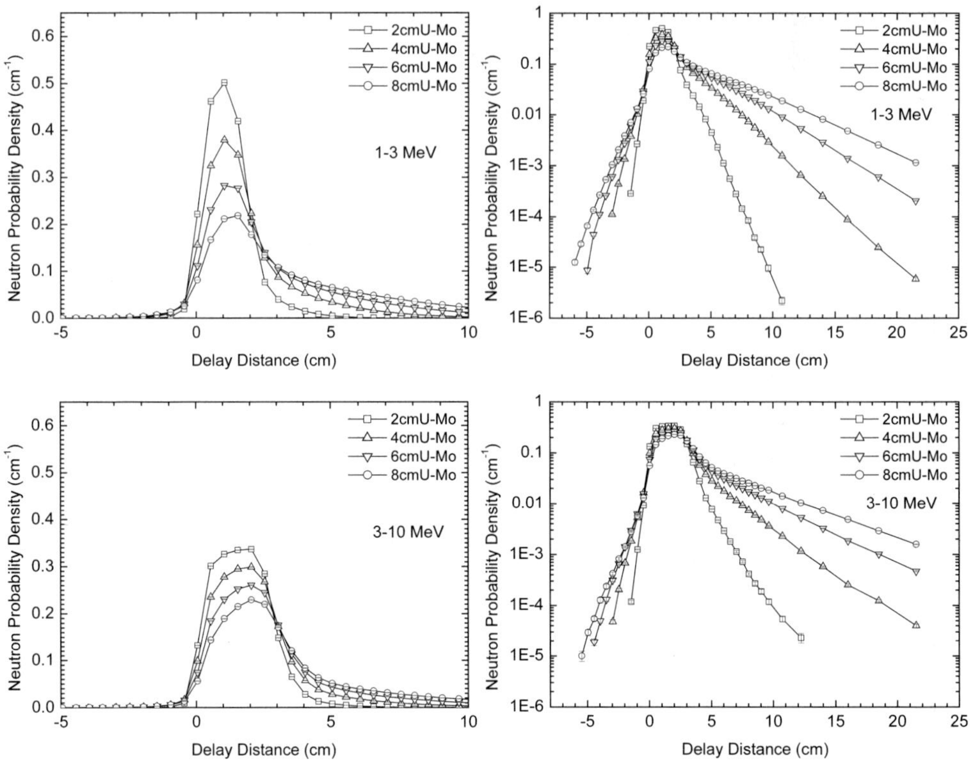

Figure 4.2. Resolution functions for U-Mo spheres of variable radius in linear (left) and logarithmic scale (right), in energy range 1-10 MeV.

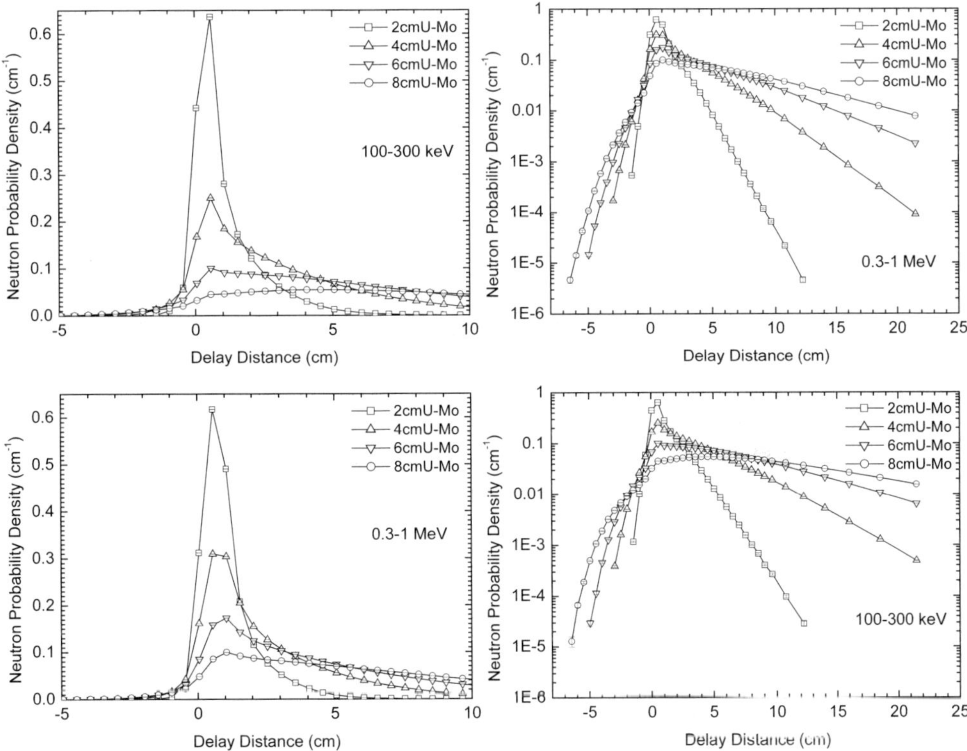

Figure 4.3. **Resolution functions for U-Mo spheres of variable radius in linear (left) and logarithmic scale (right), in energy range 0.1-1 MeV.**

4.2.2 *Target features affecting resolution functions – neutron low-energy range*

In the previous sections the high-energy range has been discussed. However, many neutron cross section measurements performed at the GELINA facility require low-energy neutrons. The direct neutron spectrum of the existing GELINA target is used only for energies above 100 keV. For neutrons of lower energies, a moderator is used. In that case only the neutrons coming from the moderator are allowed to reach the detector. The rest of the neutrons is eliminated by shielding bars placed between the target and the flight path. Therefore it is worthwhile to investigate not only the influence of the target itself on the RFs but also the influence of the moderator. In order to disclose this influence, a dedicated investigation has been carried out, of which results are presented and discussed next.

Again the U-Mo sphere has been chosen for a target. The radius of the sphere was fixed to 2 cm. Water shells with thicknesses 2, 4, 6, and 8 cm were added around the U-Mo sphere. In contrast to the investigation for the high-energy range (see section 4.2.1), an isotropic neutron source with energy of 2 MeV producing a 1-ns pulse was placed at the centre of the sphere. In this manner, the influence of the shape of the electromagnetic cascade on the RFs has been eliminated. The energy of the neutron source was inspired by the average energy of the neutrons produced in the U-Mo sphere with R = 2 cm, if the source specified in section 4.2.1 is used (1.6 MeV). A point detector was located 60 m from the sphere's centre.

Figure 4.4 shows the RF results for variable amounts of H_2O moderator. It is evident that even the addition of only 2 cm of water changes the features of the distributions significantly. In particular, the distinction between the peak and the tail region can no longer be made as a smooth shape is observed at all energies and water shell thicknesses ($\geq$ 2 cm). In fact, as can be seen for the 3-10 keV simulation the water-moderated RFs are considerably broader than those without the presence of water. However, considerably fewer neutrons would be produced at each of these energies if no water is added. In particular, it would be difficult to obtain a statistically significant result for the waterless simulation and the energy bin of 1-10 eV. So, in practice a neutron source for these energies is unrealistic without the use of an efficient moderator. The RF broadening becomes more significant with decreasing neutron energy and increasing water thickness. The shift towards more negative values of the delay distance with increasing thickness of the water shell is caused by fast neutrons that undergo their final scattering close to the surface that is nearest the detector. As the energy following this final scattering is considerably lower than the energy of the neutrons that reached this region of the water shell, the effective distance that these neutrons travel to the detector is lower than the nominal value, whereas the time of flight prior to these final interactions is negligible.

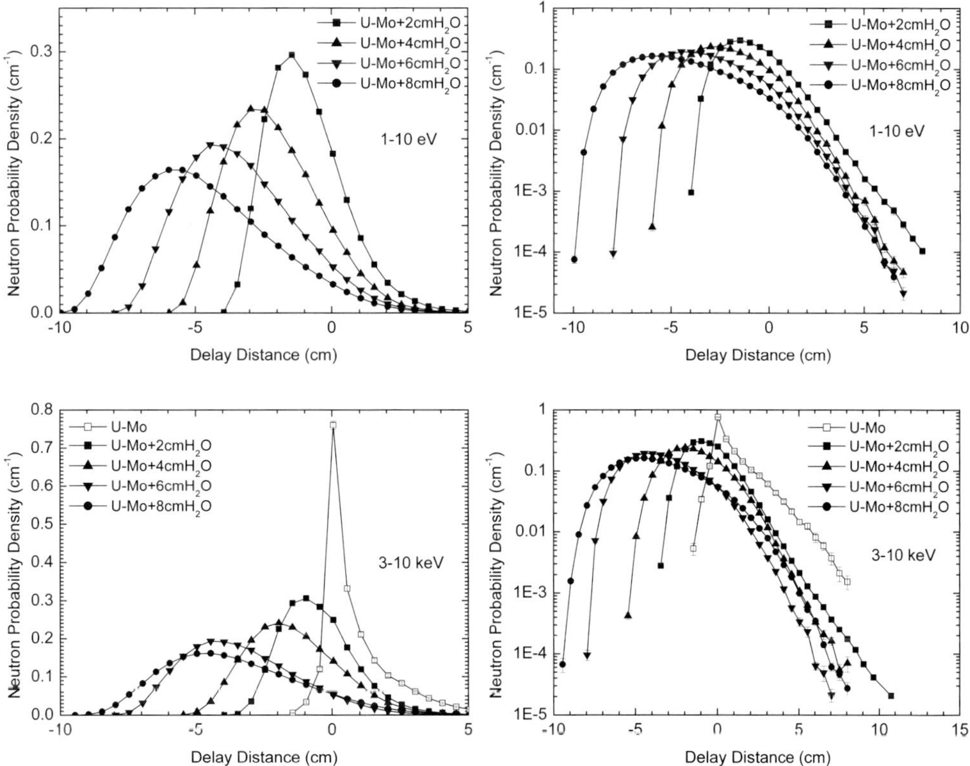

Figure 4.4. **Resolution functions for U-Mo spheres with 2-cm radius surrounded by water shells with various thicknesses, in linear (left) and logarithmic scale (right).**

4.2.3 *The resolution functions of the compact target using the existing moderator*

As was shown previously, the shape of the RFs is influenced both by the target material and the moderator. Figure 4.5 depicts the RFs for the moderated neutron spectrum of the existing GELINA target and the compact cylindrical target with H = 5 cm, R = 1.5 cm. The moderator presently used at GELINA was used (see chapter 2). The moderator tanks were placed above and below the U-Mo cylindrical target. In this simulation only the neutrons coming from the moderator could reach the detector. All neutrons traveling from the target itself towards the detector were eliminated so that none of them could contribute to the RF distributions. Naturally, this situation represents an ideal case. In reality, some neutrons can still reach the detector after being scattered by other materials than by the moderator (for example collimators). However, the probability of these events is very small. The pencil electron source introduced earlier was used in the simulations.

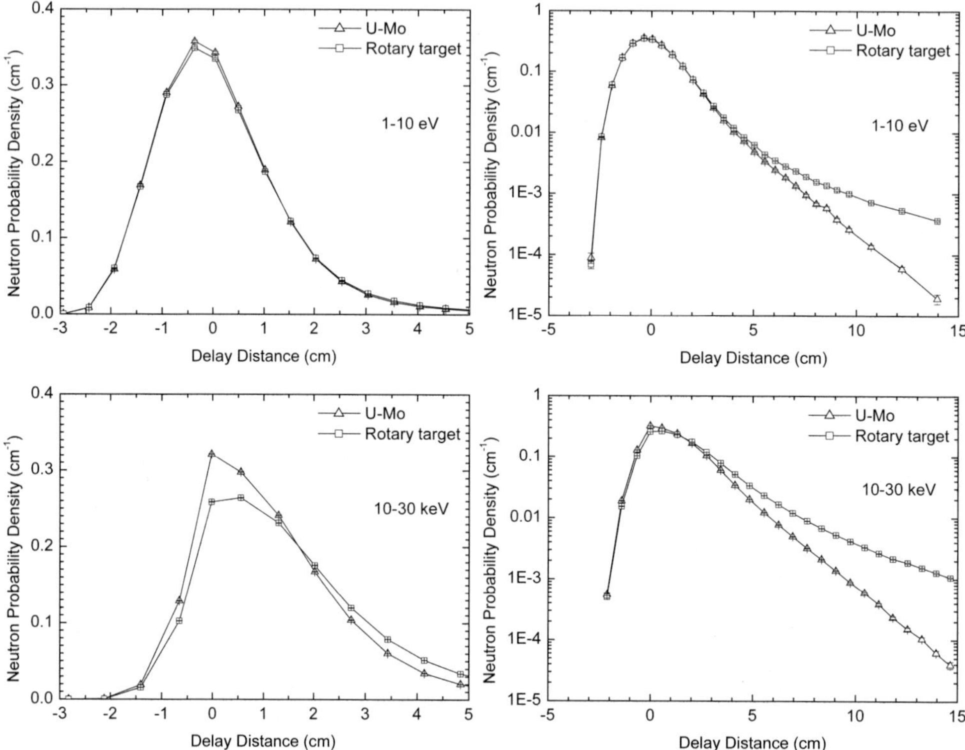

Figure 4.5. Resolution functions for the moderated neutrons at 90° for U-Mo cylinder H = 5 cm, R = 1.5 cm, and for the GELINA target using the existing moderator, in linear (left) and logarithmic scale (right).

The results pictured in Figure 4.5 demonstrate that by using a compact target, combined with the existing moderator there is no significant RF improvement for low neutron energies (see energy bin 1-10 eV). The moderator mainly dictates the shape of the RF in the peak area, and only a small difference is observed in the peak. For higher delay distance values a larger difference can be observed. However, this difference appears for values beyond FWHUM, and therefore it has a very low importance (see chapter 3, section 3.3.4). In the energy range 10-30 keV a relevant improvement can be observed in the peak and the tail of the RF. This reveals the potential for an improvement of the high-energy cross section measurements, which are relevant for high-energy nuclear systems. In addition, based on the results, it can be concluded that the target has only a minor influence on the RFs for the energies around 1 eV.

For the moderated neutron flux originating from the compact target and the existing moderator, one can expect less moderated neutrons compared to the present setup. This assessment can be made considering the discussion in chapter 5, in which it is concluded that the existing GELINA target emits more neutrons towards the moderator tanks. However, a dedicated moderator for the new target could eliminate this flux decrease in the low-energy range. This part is left for future investigation.

4.3   Conclusions - What is the way to optimize a new target?

From the results presented in this chapter it comes out that the optimal shape of a new target should be the result of a compromise. The neutron production increases with the volume of the target. The further contribution to the production becomes negligible once a certain target volume is reached. However, the need of having an excellent resolution leads to the effort to minimize the volume of the target. The study shows that a substantial reduction of the target volume, compared to the present GELINA target, is prerequisite to improve the RFs. Therefore, a dedicated figure of merit must be applied to perform a quantitative comparison of various designs. This FOM will be discussed in detail in the next chapter.

As will be shown in chapter 5, a significant improvement in the neutron flux, combined with superior RFs in the high-energy range can be achieved by using a compact target. The results from section 4.2.3, on the other hand, reveal that the target itself does not have any relevant influence on the RFs for low neutron energies. Therefore, it was decided to focus on the high-energy range, while leaving the low-energy range for later investigation. The neutronics properties of the target-moderator assembly are mainly given by the moderator properties, and therefore a separate moderator design study will be performed in the future.

Chapter 5

Towards a compact neutron producing target – neutronics point of view

---

In the previous chapter it was shown that a considerable reduction of the target volume is required to improve the RFs in the high-energy range. This can be accomplished without compromising neutron production. In the following the 4π neutron production of various materials will be presented to compare other promising materials to U-Mo, from the neutronics point of view. Afterwards, the angle-dependent neutron fluxes of cylindrical targets of various materials will be discussed and compared with those of the present rotary target. The investigation has been performed for different flight paths. Next, the angle-dependent fluxes of various compact U-Mo geometries will be discussed too. The RFs for different materials as well as for compact U-Mo designs were also compared and will be shown afterwards. Finally, in order to formulate an objective criterion for the different design possibilities, a figure of merit will be discussed and evaluated.

5.1 The neutron flux

5.1.1 The choice of the target material for a new compact target

The rotary target presently in use has been built using U-Mo alloy as an electron-neutron converter. However, it was not a priori clear that a minimum volume uranium target would withstand the maximum GELINA electron-beam power of about 10 kW, if this material is chosen. Therefore, alternative target materials attractive from a thermal and neutronics point of view were investigated as well. The considered materials were tantalum (Ta), tungsten (W) and thorium (Th). In addition, Hg as a potential coolant was investigated too. In the investigation all target cylinders have equal dimensions in units of the radiation length X_0 and Moliere radius R_M.

For the chosen materials the corresponding X_0, R_M and E_C values are given in Table 5.1. It has been concluded previously that the U-Mo cylinder with H = 5 cm, and R = 1.5 cm provides roughly the same amount of neutrons as the present rotary target. This corresponds to H = $16.234 X_0$ and R = $1.518 R_M$. Therefore, the investigation of a new design started from this particular geometry and the same electron-beam characteristics as before. To simplify the calculation of X_0 the presence of Mo in the U-Mo alloy was neglected by assuming A = 238 (see Equation 4.5 in chapter 4). From the comparison in Table 5.1 it is apparent that the use of other material than U will cause about 45% loss in the 4π neutron production ($4\pi Y_n$). Only in the case of Th this loss is smaller (20%). Clearly, U is the best choice for the target material to maximize the neutron production. In case another material than U is used, a significantly lower neutron yield can be expected.

Material	*Z*	*A*	ρ *(g/cm³)*	X_0 *(cm)*	R_M *(cm)*	E_C *(MeV)*	*H (cm)*	*R (cm)*	$4\pi Y_n$ *(% n/e)*
Ta	73	180.95	16.65	0.410	1.048	8.22	6.66	1.59	3.39
W	74	183.84	19.25	0.351	0.910	8.11	5.70	1.38	3.37
Hg	80	200.59	13.55	0.472	1.320	7.51	7.66	2.00	3.11
Th	90	232.04	11.72	0.508	1.595	6.69	8.25	2.42	4.79
U	92	238.00	19.05	0.308	0.988	6.54	5.00	1.50	5.93

Table 5.1. **Size of the cylindrical targets of various materials calculated by using the scaling with radiation length and Moliere radius.**

5.1.2   *Angle-dependent flux*

Until this point, the discussion was focused on the 4π neutron production. However, the flux level at a given flight path is of the user's interest at GELINA as it directly influences the time necessary to obtain the required accuracy of the measurement. Accordingly, this issue has been examined and is presented next. An average electron current of 100 µA has been used for the postprocessing of the simulation results.

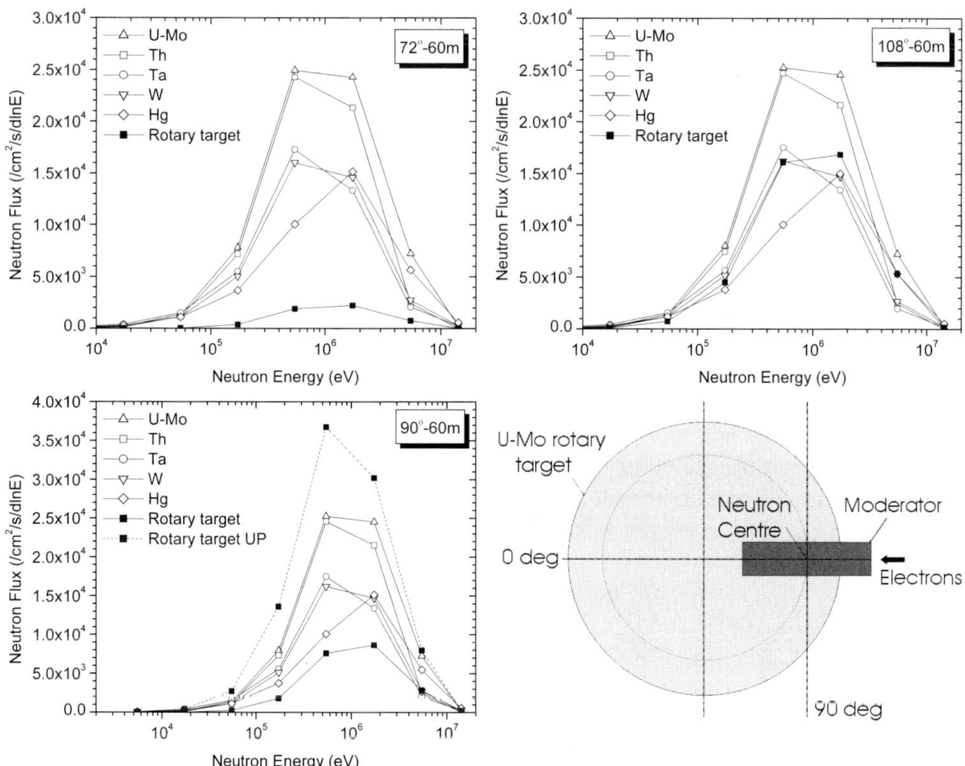

Figure 5.1. **Neutron flux per unit lethargy at 72°, 90° and 108° for cylindrical targets of materials U-Mo, Ta, W, Th, Hg, and for the GELINA rotary target; relative errors < 5%. Top view of the existing GELINA target (lower right) shows the centre of the neutron production, which is located on the axis of the moderator tanks.**

Simulations of the angle-dependent neutron flux were carried out [Fla05]. In Figure 5.1 a comparison is given of the neutron flux per unit lethargy for a cylindrical target of different materials, and for angles of 72°, 90°, and 108° with respect to the electron-beam direction. The detector was placed at 60 m in the simulation, and the usual GELINA electron source has been applied. In the case of the angle of 72°, the flux of the compact U-Mo target is higher by a factor of 12 at the maximum, compared with the existing target. The reader can realize that in the rotary target there is a large amount of material between the detector and the neutron-creation volume due to the large size of the target. The diameter of the rotary target is 32.4 cm, while the theoretical neutron-creation centre is located 12.4 cm from the rotation axis (see drawing in Figure 5.1). The thickness of material between the neutron-creation centre and the detector varies with the angle of the detector location. It is 14.9 cm, 10.4 cm, and 7.3 cm for 72°, 90°, and 108°, respectively. This heavy material scatters many high-energy neutrons away from the track towards the detector. Therefore, the neutron flux changes very rapidly with angle, as shown in Figure 5.1.

Since many neutrons are scattered, the flux of the rotary target is much higher in the upward and the downward direction than in the horizontal direction (see the angle of 90°, curve 'Rotary target UP'). This fact justifies the choice to place the present moderator tanks above and below the GELINA target to produce a high number of low-energy neutrons. In the case of the angle of 108°, the amount of material between the detector and the theoretical neutron-creation point is much lower than for 72°. Still, the neutron flux of the U-Mo cylinder is higher by a factor of 1.5 at the maximum. Based on these results, it can be concluded that the U-Mo alloy is the best material of choice from the neutronics point of view. This conclusion is consistent with the conclusion from the previous section. Finally, it should be stressed that the cylindrical geometry represents a much more isotropic neutron source than the rotary GELINA target. Therefore, such a compact target would further enhance the versatility of the GELINA facility as it can provide various flight paths with similar neutron fluxes.

One should realize that Hg also performs very well in the simulations discussed above. This is an important remark as this material is intended as a coolant of the new target design. The use of heavy material for target cooling arises from one of the project conditions, which requires producing fast and moderated neutrons separately in the new target. As mentioned earlier the research presented here has been focused on the optimization of the target in the high-energy range. The low-energy range will be subject of future work as that problem is related to the shape and composition of the moderator.

5.1.3   Compact U-Mo neutron producing targets

As concluded in the previous section, the optimal material to obtain a large number of neutrons is the U-Mo alloy, which is also the material used to build the existing GELINA target. The optimal shape of the new target has to be inspired by the shape of the electron-photon cascade as in this way the neutron escape can be maximized without excessive use of material. Therefore, as a first step, a cylindrical geometry made of U-Mo alloy was chosen for the initial investigation. However, from an engineering point of view, it is envisaged that right-angled plates will be a more reasonable choice for a new target design. Accordingly, simple block-shaped geometries were investigated too, and the further optimization has been based on this type of design. The size of the initial U-Mo block was inspired by the U-Mo cylinder with H = 5 cm and R = 1.5 cm, resulting in a size of 3 x 3 x 5 cm³.

In this section six different target designs are compared from a neutronics point of view. Besides the presently used rotary target, the following geometries were simulated and tested:

- U-Mo cylindrical target with R = 1.5 cm, H = 5 cm,
- U-Mo block of 3 x 3 x 5 cm³,
- U-Mo block of 4 x 4 x 5 cm³,
- Initial target design consisting of ten 3 x 3 cm² U-Mo plates with thicknesses of 0.2 cm, 5 x 0.15 cm, 0.2 cm, 0.3 cm, 0.5 cm, and 2.15 cm. Eleven 1-mm-Hg cooling channels were placed between the plates. The geometry starts with a Hg channel so that the first plate of the geometry is cooled from both sides, like the rest of the plates (see drawing in chapter 6).
- Final target design consisting of seven 3 x 3 cm² U-Mo plates with thicknesses of 0.15 cm, 2 x 0.2 cm, 0.3 cm, 0.45 cm, 0.9 cm, and 2.1 cm. Each U-Mo plate is fully coated by Ta cladding with a thickness of 0.2 mm. Seven 1-mm-Hg cooling channels were positioned between the plates. In this case, the geometry starts with the first U-Mo plate, e.g., this plate is cooled from backside only (see drawing in chapter 6).

The optimization procedure is discussed in detail in chapter 6, which concerns the heat-removal aspects of the design study. In addition, also an explanation is given why the final target design described above is considered as the optimal geometry.

Figure 5.2 presents the neutron flux per unit lethargy for these geometries at angles of 72°, 90°, and 108°. The detector was placed at 60 m, and the usual GELINA electron source has been used in the simulations. In the case of the angle of 72°, the neutron flux of the final design is higher by a factor of 12.1 at maximum, compared with the existing rotary target. Again, this comes from the fact that in the rotary target there is a large amount of material between the detector and the volume of neutron creation. In the case of the angle of 108°, the amount of material between the detector and the neutron-creation point is much lower than for 72°. Still, the neutron flux of the final design is higher by a factor of 1.4 at maximum when compared with the rotary target.

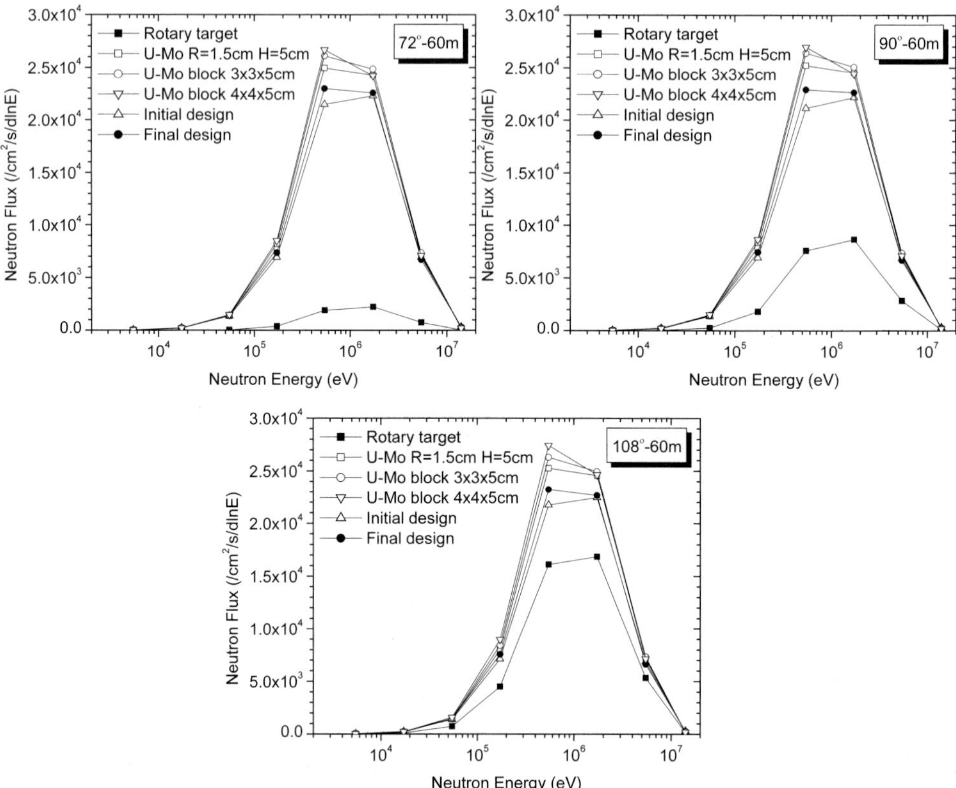

Figure 5.2. **Neutron flux per unit lethargy at 72°, 90°, and 108° as a function of the neutron energy for various target designs, with relative errors < 5%.**

From the comparison of the U-Mo block geometries of 4 x 4 x 5 cm³ and of 3 x 3 x 5 cm³ it appears that the energy-integrated neutron flux is almost the same in both cases. Therefore, to use the least target material possible to optimize the RFs, the size of 3 x 3 cm² was chosen for further investigation. The design of 3 x 3 x 5 cm³ leads to a flux higher by 3.7%, compared to the cylindrical target. This small difference is caused by additional material of the block-shaped target, which contributes to the neutron escape (see discussions in chapter 4).

The first phase change of U appears at 668 °C, therefore a temperature limit of 450 °C was chosen as a conservative approach (see discussion in chapter 6). Since maximum temperatures of a single-block U-Mo target would easily exceed 450 °C for a 10-kW electron beam, a sliced design of 3 x 3 cm² has been proposed instead, with well-specified thickness of each plate. Subsequently, the target optimization has been carried out, starting from U-Mo block of 3 x 3 x 5 cm³ to maximize the neutron production, while keeping the maximum plate temperatures below the limit. This has been done by minimizing the amount of Hg and maximizing the thickness of each U-Mo plate. See chapter 6 for more details about the optimization procedure.

From the comparison of the initial design versus the final design in Figure 5.2 it is apparent that no significant improvement could be obtained by additional optimization. An improvement of only 4.5% in the flux at 90° has been achieved in this case. The final design provides the flux lower by 10.9% when compared to the U-Mo block of 3 x 3 x 5 cm³. This difference is caused by the lower amount of U-Mo alloy in the final design. It is also important to stress that the final design represents a more isotropic neutron source than the GELINA target and thus can provide many flight paths with similar neutron fluxes. This statement is justified by flux results for different angles.

5.2   Resolution functions

5.2.1   *The comparison of alternative target materials*

In order to compare the RFs of the compact cylindrical geometries described in Table 5.1 with those of the present target, simulations were carried out in which the 1-ns electron pulse bombarded the cylinders of the various materials. The electron pulse had an identical energy distribution as earlier, e.g. the uniform energy range from 70 to 140 MeV. Equivalently to the neutron-flux comparison, the considered materials are Ta, W, Th and Hg. The usual U-Mo alloy has been used instead of pure U mentioned in Table 5.1.

As shown in Figures 5.3 and 5.4, the compact targets are better in the RFs owing to the smaller amount of material forming the targets. The results are presented both in a linear scale and in a logarithmic scale so that the differences in the tail of the RFs are visible too.

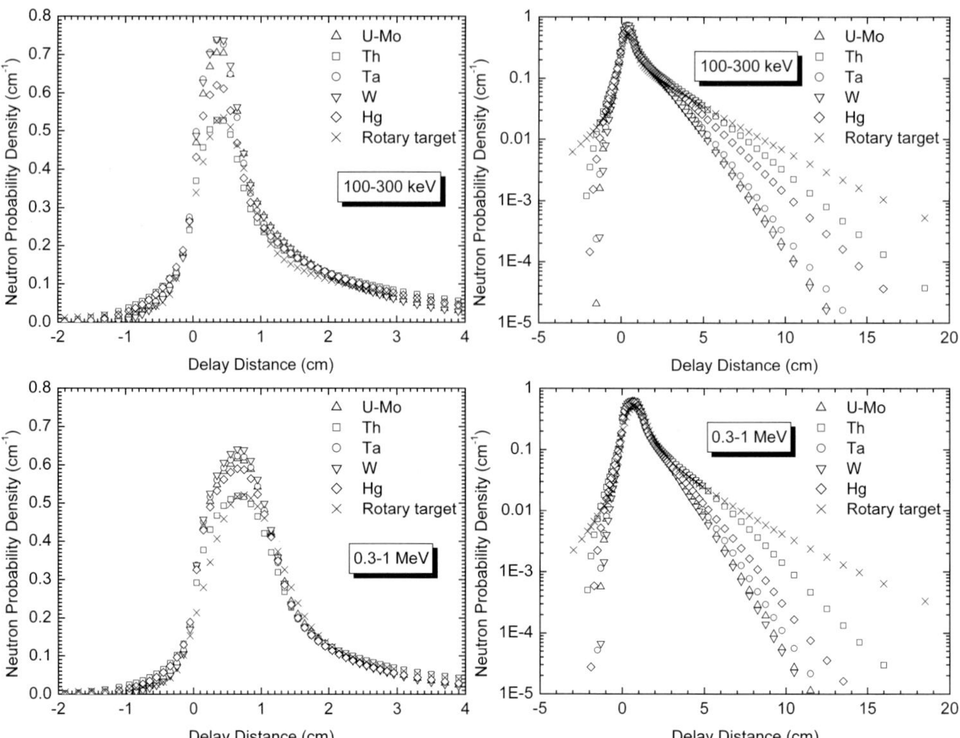

Figure 5.3. Resolution functions at 90° for cylinders of U-Mo, Ta, W, Th and Hg and for the GELINA rotary target in linear (left) and logarithmic scale (right), in energy range 0.1-1 MeV. Relative errors < 5%.

Clearly, the differences increase with decreasing neutron energy. The use of Ta and W gives results comparable with U-Mo. Evidently, both these materials can be reasonable substitutes for U-Mo, if the quality of the RFs is the only condition. The RFs of the Th target, on the other hand, are significantly broader than for the rest of materials. It is interesting to observe that the RFs for Th have similar shape when compared with the rotary target. For neutron energies above 1 MeV, in the case of the rotary target, a slight shift towards positive delay distance values is apparent from Figures 5.3 and 5.4, compared to compact designs. The size of the rotary target causes this shift. It is obvious that high-energy neutrons undergo only few collisions (or none) before reaching the detector. Still, the high-energy neutrons undergo on average more collisions in the present target than in the cylindrical target. This is reflected in more positive delay distance values.

At high energies, a significant broadening of the RFs is observed in the results. The 1-ns electron pulse causes this broadening. Finally, it can be stated that no significant improvement of the RFs can be reached for neutron energies above 3 MeV. In such a case, the improvement in the RF tail is of low importance as it occurs only for values below the Full Width at one Tenth of the Maximum (FWTM).

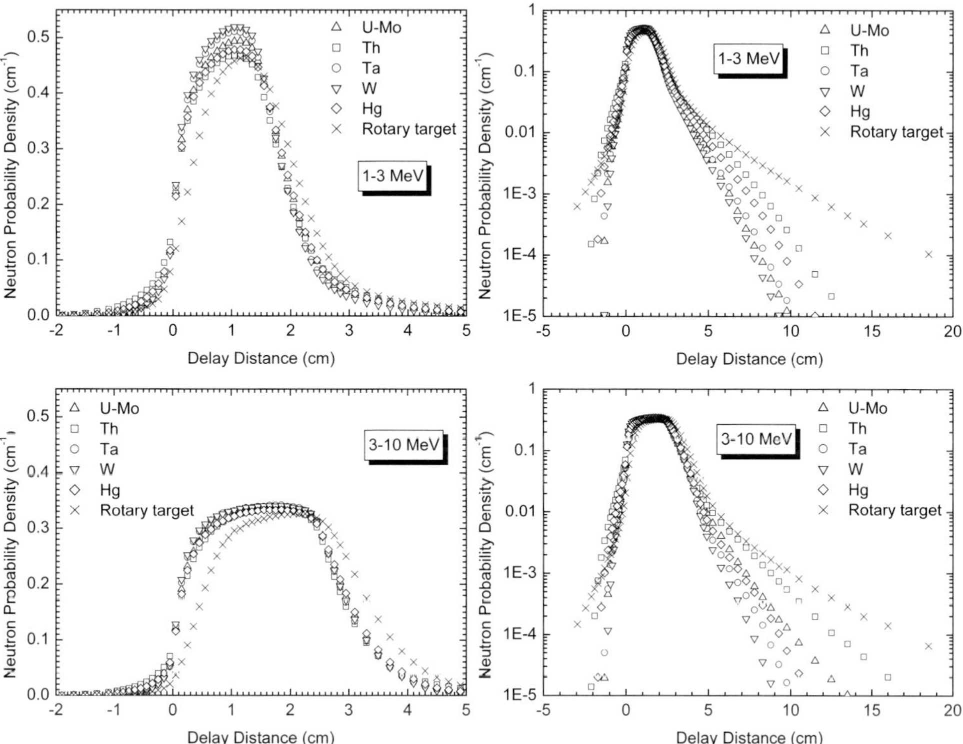

Figure 5.4. Resolution functions at 90° for cylinders of U-Mo, Ta, W, Th and Hg and for the GELINA rotary target in linear (left) and logarithmic scale (right), in energy range 1-10 MeV. Relative errors < 5%.

5.2.2   Compact U-Mo neutron producing targets

Similarly to the investigation described in section 5.1.3, five compact U-Mo target designs, including the final design were compared with the rotary target. As before, the RFs represent the response of the targets on a 1-ns electron pulse with a uniform energy distribution between 70 and 140 MeV.

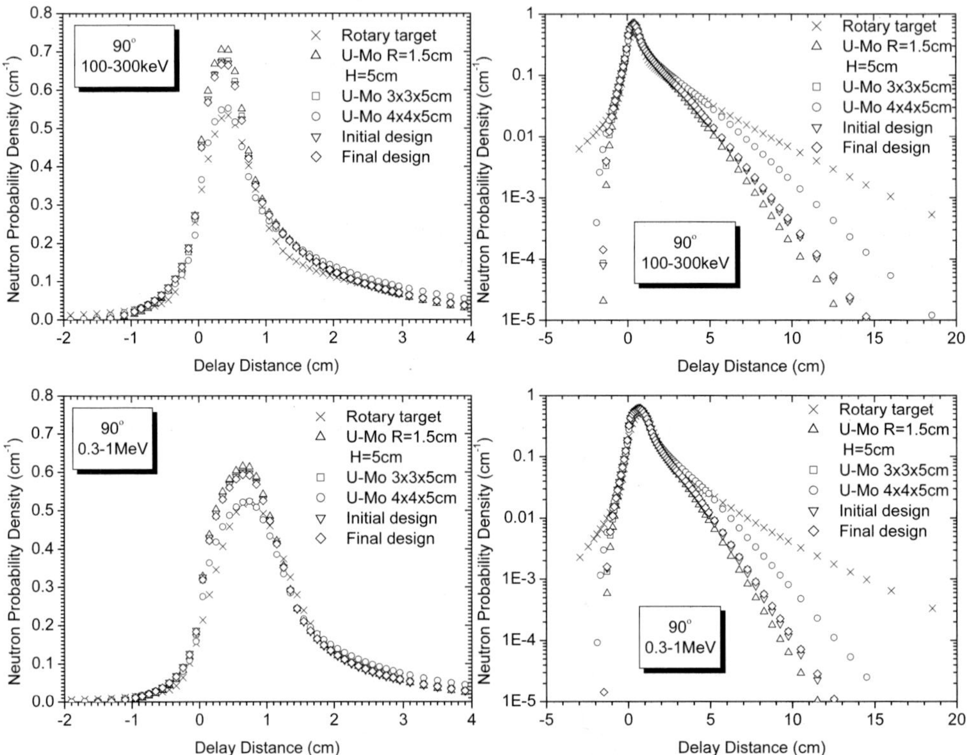

Figure 5.5.   Resolution functions at 90° for various target designs in linear (left) and logarithmic scale (right), in energy range 0.1-1 MeV. Relative errors < 5%.

Figures 5.5 and 5.6 present the comparison of the RFs at 90° for different energy bins. The results are again presented both in a linear scale and in a logarithmic scale. It can be seen in Figures 5.5 and 5.6 that the compact targets are superior to the rotary target. In all cases, the U-Mo cylindrical target produces the narrowest RFs. All other compact geometries also perform very well. The only exception is the U-Mo block of 4 x 4 x 5 cm³, which causes larger broadening of the RFs due to its size when compared to other compact designs. However, the differences in RFs decrease with increasing neutron energy, and therefore no significant improvement of the RFs can be reached for neutron energies above 3 MeV.

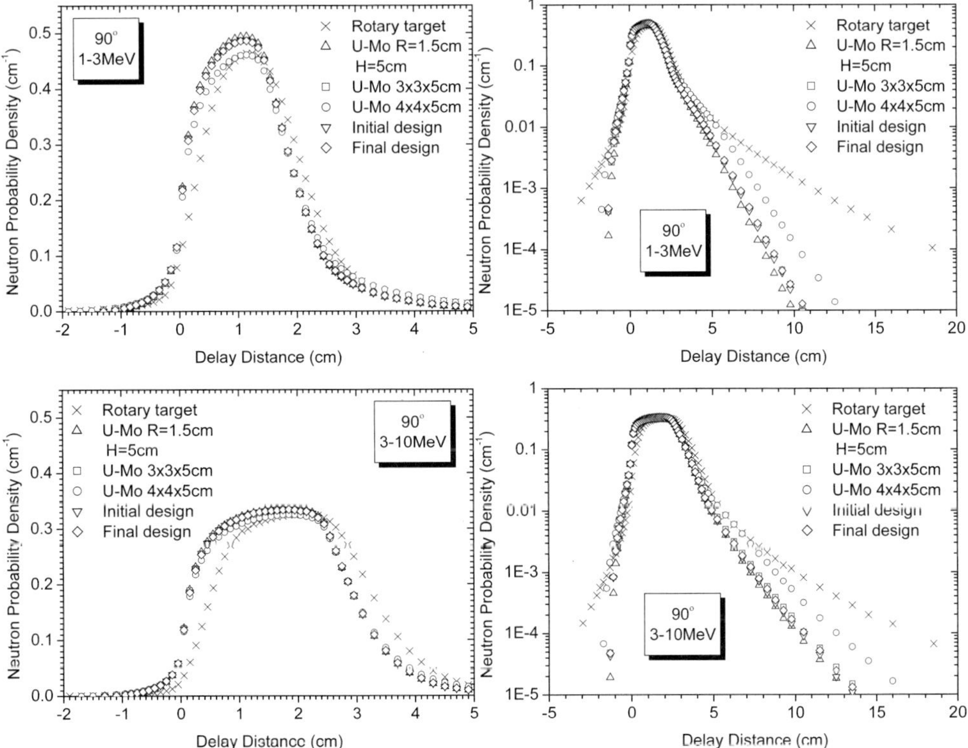

Figure 5.6. **Resolution functions at 90° for various target designs in linear (left) and logarithmic scale (right), in energy range 1-10 MeV. Relative errors < 5%.**

5.3   Figure of merit

It was suggested in the past that the quality of different TOF facilities can be assessed by a **Figure Of Merit** (FOM) [Rae70], which is defined as the average neutron flux at the end of the flight distance necessary to obtain a given energy resolution. In general, this FOM approach presupposes that one can select the flight path length at will, if a particular resolution is required. In practice, the choice of flight path length depends on other aspects, such as prompt-γ-flash interference, the availability of flight path lengths, and the flexibility of the measurement equipment.

In order to carry out a comparison between promising GELINA designs, and to compare the neutron-beam quality at different angles, an energy-dependent FOM is introduced in this section, which follows the suggestion above.

Generally, the neutron flux Φ (in m²s⁻¹) of any facility at a given distance L (in m) from the neutron source can be expressed as

$$\Phi = \frac{I_n}{L^2} \tag{5.1}$$

where I_n is the neutron source intensity (in s⁻¹). To calculate a hypothetical flight path length L to obtain a prefixed energy resolution, Equation 1.3 from chapter 1 can be modified to the following expression:

$$L = \frac{2}{\frac{\delta E_n}{E_n}}\sqrt{\delta L^2 + 1.9 E_n \delta t^2} \tag{5.2}$$

Combining Equations 5.1 and 5.2 we obtain

$$\Phi(E_n) = \frac{\left(\dfrac{\delta E_n}{E_n}\right)^2}{4} \frac{I_n}{\delta L^2 + 1.9 E_n \delta t^2} \tag{5.3}$$

The first fraction in Equation 5.3 is constant for a fixed energy resolution. Consequently, the second fraction is assumed to represent a FOM of an experimental facility

$$FOM(E_n) = \frac{I_n}{\delta L^2 + 1.9 E_n \delta t^2} \tag{5.4}$$

In order to calculate such a FOM with the MCNP code for various target designs, Equation 5.4 was rewritten as follows:

$$FOM(E_n) = \frac{d\phi_{MCNP}}{d\ln E} L_{sim}^2 \frac{1}{\left(FWHM_L\right)^2 + 1.9 E_{center} \left(FWHM_t\right)^2} \tag{5.5}$$

where $d\phi_{MCNP}/d\ln E$ (in cm^{-2}s^{-1}) is the MCNP neutron flux per unit lethargy at the simulation distance L_{sim} (in cm) of the MCNP detector, $FWHM_L$ is the full-width-half-maximum value (in cm) of the RF for given energy bin, E_{center} is the energy centroid (in eV) of the energy bin, and $FWHM_t$ is the full-width-half-maximum value (in µs) of the initial electron pulse. For the neutron transport related part of the energy resolution, the FWHM is taken [Coc02]. However, because some high-resolution experiments can be affected by the resolution-function asymmetry, e.g. by the RF tails, also a FOM based on the FWTM is included in the assessment. The FOM specified in Equation 5.5 is a modification of the FOM used in the past [Coc02] for the GELINA facility. The modification concerns the neutron energy spectrum (the previously used FOM considered the moderated neutrons) as well as the possibility to use the point-detector flux simulations instead of using the angle bins. The postprocessing of the simulation data has been done for an average electron current of 100 µA, with a constant value of $FWHM_t = 0.67$ ns.

5.3.1   *The comparison of alternative target materials*

The FOM was calculated for the cylindrical targets with materials. The results are depicted in Figure 5.7. The angles 72°, 90°, and 108° are presented not only for $FWHM_L$, but also for $FWTM_L$ of the RF in given energy bin. The $FWTM_L$ value carries information about the RF asymmetry, which can play an important role for some high resolution experiments.

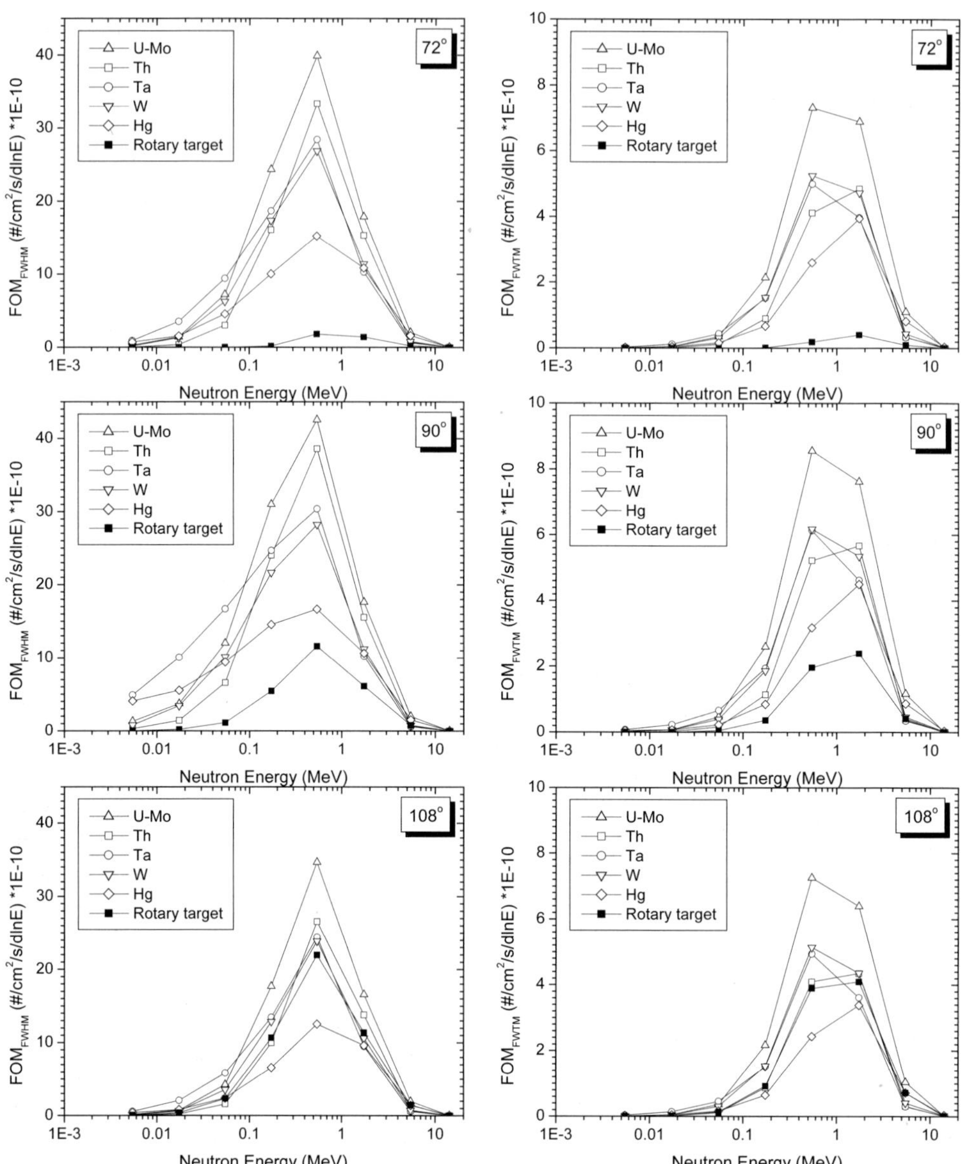

Figure 5.7. Figures of merit for FWHM (left) and for FWTM (right) values at 72°, 90°, and 108° for U-Mo, Ta, W, Th and Hg cylinders and for the GELINA rotary target, multiplied by a factor of 1E-10. Relative errors < 5%.

The results from Figure 5.7 represent a combination of the neutron flux results and the RF results described earlier. For 72°, the FOM_{FWHM} for the U-Mo cylinder is higher by a factor of 20 at the maximum when compared with the rotary target. This difference is mainly due to the flux difference of given geometries. For 90°, the FOM_{FWHM} difference decreases to a factor of 4, due to the relative increase of the neutron flux of the present target. The results for 108° again reveal a strong influence of the flux on the FOM, however, the value of the existing GELINA target is lower only by a factor of 1.6 compared to the U-Mo cylinder. The results for the FOM_{FWTM} are similar in relative numbers to the FWHM values, although for some neutron energies even a higher FOM improvement is achieved. This is caused by a larger amount of U-Mo of the rotary target, which broadens the RFs, especially in the tail.

5.3.2   Compact U-Mo neutron producing targets

In order to assess the quality of various U-Mo designs, a comparison of FOMs at angles of 72°, 90°, and 108° has been carried out. The results were obtained using the fluxes and RFs introduced earlier in this chapter. The FOM values were calculated from Equation 5.5 for both FWHM and FWTM.

The results are shown in Figure 5.8. Generally, the FOM results are in close relation with the neutron flux, which is a major component of the FOM. At 72°, the FOM_{FWHM} for the final design is higher by a factor of 20 at maximum, compared with the rotary target. At 108°, this factor decreases to 1.4. It can be observed in Figure 5.8 that the FOMs for the U-Mo block of 4 x 4 x 5 cm^3 are lower at maximum when compared to the block of 3 x 3 x 5 cm^3. Especially for FOM_{FWTM} this difference is large. This is because of the material at the periphery of the block of 4 x 4 x 5 cm^3, which does not significantly contribute to the neutron production. Instead, it broadens the RFs, which has the direct impact on the FOM. The reader can realize that the results for the FOM_{FWTM} are similar in relative numbers to the FOM_{FWHM} values. Moreover, a higher FOM improvement at maximum is achieved, with factors of 33 and 1.5 at angles of 72° and 108°, respectively. This is caused by a much larger amount of U-Mo of the rotary target, which causes an additional broadening of the RFs.

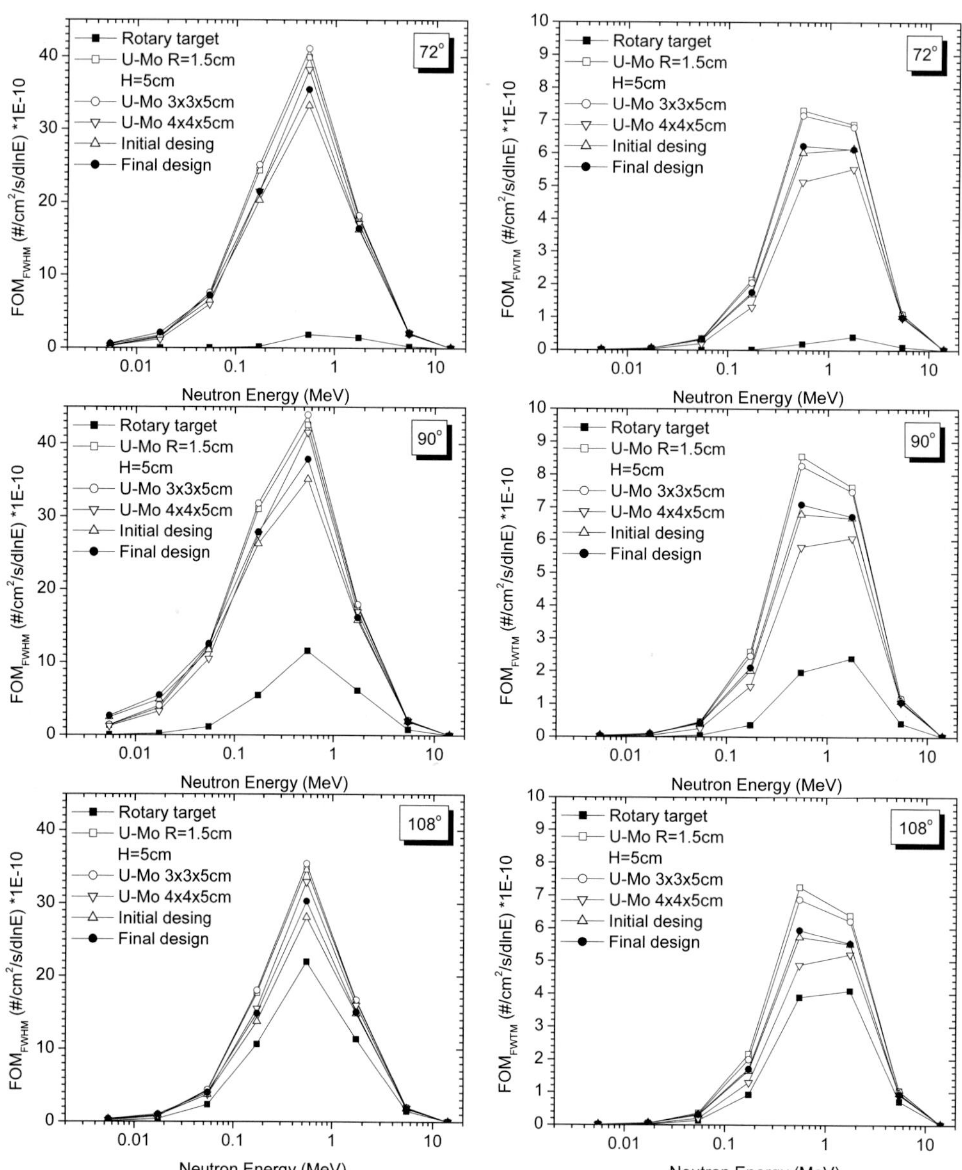

Figure 5.8. Figures of merit for FWHM (left) and for FWTM (right) values at 72°, 90°, and 108° for various U-Mo targets and for the GELINA rotary target. Results multiplied by a factor of 1E-10, with relative errors < 5%.

From the comparison of the initial design versus the final design in Figure 5.8 it is apparent that no relevant improvement could be obtained by additional optimization. Improvements of 6.6%, 7.8%, and 7.8% have been achieved in the FOMs (FWHM) at maximum for angles of 72°, 90°, and 108°, respectively. A detailed description of the optimization procedure is given in chapter 6.

5.4   Conclusions

In this chapter it is shown that the use of other material than U for a new neutron producing target would cause about 45% loss in the 4π neutron yield. Only in the case of Th this loss is 20%. Therefore it can be concluded that U is the best choice for the target material to maximize the neutron yield. A compact U-Mo target produces the neutron flux at 72° higher by a factor of 12 at maximum, compared with the present rotary target. In addition, this material also performs very well from the resolution function point of view. This statement is fully confirmed by the FOM results, from which it is apparent that U is the best material to maximize the neutron yield, while producing very narrow RFs. The FOM_{FWHM} for the U-Mo cylinder at 72° is higher by a factor of 20 at the maximum, compared with the rotary target. This difference is mainly caused by the flux difference of these geometries. For 90°, this difference decreases to a factor of 4, as a consequence of the flux increase of the rotary target. The final design, as proposed by interrelated neutronics and heat transfer simulations, has the FOM_{FWHM} at 72° higher by a factor of 20 at maximum, compared with the rotary target. It also has been realized that the final design, unlike the GELINA rotary target, can provide various flight paths with similar neutron fluxes. This would be another asset of using the compact target as a neutron source.

Chapter 6

Towards a compact neutron producing target – heat transfer point of view

The neutronics properties of the compact targets have been proven to be superior to the existing rotary target. It also has been realized that uranium is the optimal target material to maximize the neutron production. The use of a different material than uranium would lead to a substantial loss in the neutron production. U-Mo alloy was chosen instead of pure metallic uranium for the present rotary design to improve the mechanical properties of the target material. Consequently, a decision was taken to use this alloy also in a new target. This is justified by the observation that the U-Mo alloy does not significantly compromise the neutronics properties, compared to metallic uranium.

The present rotary target was deliberately designed as a massive disk for easy dissipation of the beam energy in the material (10 kW). In this way possible temperature problems were suppressed. Presently, the maximum temperature in the rotary target does not exceed 250 °C for 100 μA, which is the maximum value for the average electron current reachable at GELINA. This temperature is well below the first phase change of uranium (668 °C), at which dimensional changes appear. Such a process would lead to damaging of the target.

To accomplish the project goal of having improved RFs, the new GELINA target must be much smaller than the rotary design. The optimal size of the U-Mo geometry was discussed previously. Its main feature is that it maximizes the neutron yield without excessive use of material. However, with the decrease in size a heat transfer problem arises, possibly leading to unacceptably high target temperatures.

In the following text, an explanation is given of how the heat transfer problem has been dealt with for a compact U-Mo geometry. A combination of analytical heat transfer calculations and **C**omputational **F**luid **D**ynamics (CFD) simulations, performed with the FLUENT code [Flu01] was applied to the problem in order to optimize the geometry with respect to heat transfer.

6.1   Initial target optimization

The MCNP calculations revealed that the 4π neutron yield of the present target can be fully maintained by a cylinder with R = 1.5 cm, and H = 5 cm (see section 4.1.3 in chapter 4). Such a cylinder has a volume $V = \pi R^2 H = 35.3$ cm³. In order to reveal the temperature distribution in such a single-piece target, analytical heat removal calculations were carried out. To simplify the problem, the following assumptions were made:

- the heat generation rate per unit of volume q''' is uniform and corresponds to a total power $Q = q'''V = 10$ kW, which is the nominal power of the GELINA beam,
- the heat produced in the geometry is transported to the cylinder surface, which is kept at a constant temperature T_0,
- the steady-state condition is considered,
- the temperature distribution $T(r)$ in the cylinder is assumed to depend on the radial coordinate r only. This is not very accurate for the relatively small H/R ratio of the cylinder under consideration, but it is a conservative assumption since it will lead to an overprediction of the maximum temperature in the cylinder.

With these assumptions, the temperature distribution is given by the equation [Bee00]

$$T(r) - T_0 = \frac{q''' R^2}{4\lambda}\left(1 - \frac{r^2}{R^2}\right)$$

(6.1)

The maximum temperature that occurs at the center ($r = 0$) is then given by:

$$T_{max} = \frac{q''' R^2}{4\lambda} + T_0$$

(6.2)

For λ = 25 Wm^{-1}K^{-1}, q''' = 283 x 10^6 Wm^{-3}, and T_0 = 300 K, T_{max} = 664 °C. The first phase change of U-metal appears at 668 °C, and an upper temperature limit of 450 °C was chosen for the target as a conservative approach. The analytically calculated maximum temperature is well above this limit. In addition, this estimate of the maximum temperature assumes a uniform distribution of q''' throughout the geometry. It will be shown later in this chapter that the heat production in the target material is very inhomogeneous, and therefore, in practice, much higher temperatures would be reached. Thus, it can be concluded that the maximum temperature of a single-piece U-Mo target would exceed by far the temperature limit of 450 °C and in fact may even reach values higher than T_{max}. Therefore the investigation was focused on the optimization of a compact target with respect to reducing its maximum temperature. In particular, the possibility of segmentation of the U-Mo target into plates perpendicular to the target axis was investigated. Such a segmentation would create cooling channels in between the plates, through which a cooling liquid can flow. It was realized that, from an engineering point of view, rectangular plates are favorable over cylindrical plates, and that this has minimum impact on the neutronics properties (see chapter 5). Moreover, based on the neutronics results, it was concluded that Hg is the optimal choice for the coolant. However, Hg cannot compete with the neutronics properties of U. The most important condition for the optimization was therefore to minimize the amount of Hg between the U-Mo plates, as this has a negative influence on the neutronics properties of the design, compared to pure U-Mo. In this manner, a final target design could be obtained.

6.1.1   Segmentation of the compact U-Mo design

As was mentioned before, a U-Mo block of 3 x 3 x 5 cm^3 was chosen as a starting point for designing a new target, based on the initial cylinder with R = 1.5 cm, and H = 5 cm. In a first optimization effort, the U-Mo block was split into ten 3 x 3 cm^2 plates of varying thickness in such a way that the estimated total dissipated power per plate was below 1 kW. The thicknesses of the plates are 0.2 cm, 5 x 0.15 cm, 0.2 cm, 0.3 cm, 0.5 cm, and 2.15 cm, respectively. This geometry is later referred to as the initial target design. Eleven 1-mm Hg cooling channels are placed between the plates and before the first and after the last plate. A channel thickness of 1 mm has been chosen as an optimal choice, which was kept constant for all further optimization steps. It has to be stressed that this segmentation approach does not optimize the neutron production, but instead distributes the dissipated power quite uniformly over the plates. As a result, this design leads to a configuration with more plates than necessary, since the thickness of the plates was not optimized. Consequently, also an overestimation occurred of the number of cooling channels necessary to keep the maximum temperatures of the target below the limit.

Therefore another optimization study was carried out, aimed at finding the minimum number of Hg cooling channels and the maximum thickness for each of the U-Mo plates.

6.1.2   Plate by plate optimization

The optimization started by placing a U-Mo plate at the position in which the neutron production, as found from MCNP calculations for a solid target, is the highest (see figure 6.1). From the energy dissipation value at this location, the maximum thickness of a plate at this location was estimated, such that its temperature does not exceed 450 °C. Subsequently, additional plates were placed one by one on both sides of the first. For each plate the maximum thickness was estimated based on the energy dissipation at that location and the maximum temperature allowed.

An assumption in the above procedure is that the neutron production and the energy dissipation distribution are not strongly modified when individual plates replace the solid target. To check this assumption, two cases were studied using MCNP: the first was a 3 x 3 x 5-cm³ U-Mo block made of 1-mm plates attached to each other; the second considered twenty five 1-mm U-Mo plates, separated by the same number of 1-mm Hg channels. The latter geometry started with a U-Mo plate and ended with a Hg channel. This geometry represents an extreme situation, in which very narrow U-Mo plates were used in combination with many Hg channels. In both cases the total target length was identical. Both production distributions, as shown in Figure 6.1, closely follow a development of the electromagnetic cascade initiated in the material by the nominal GELINA electron beam.

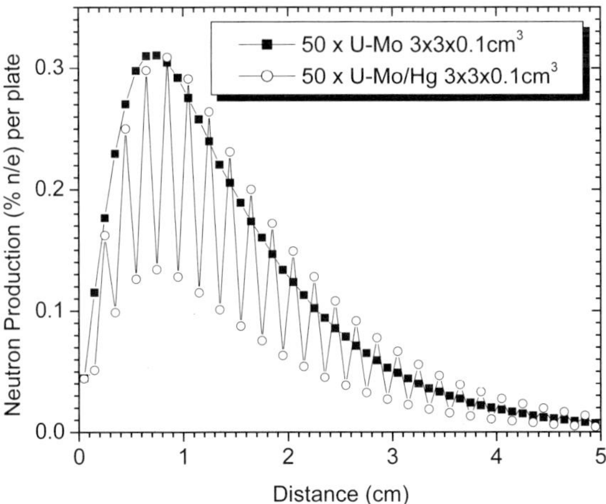

Figure 6.1. Neutron production in block-shaped geometries. The maximum neutron production is reached below 1-cm depth in both cases; these cases represent two extreme situations. The total length of the targets is 5 cm.

These two cases were chosen to investigate a possible shift in distance of the neutron production peak if a large amount of Hg is present in the target. There is a significant (22%) decrease in the 4π neutron production (5.65% n/e versus 4.41% n/c). The reason of this difference is that in the second case 50% of the U-Mo material was substituted by Hg. It has to be emphasised that the same amount of Hg would cause a somewhat different loss of neutrons, if placed at different positions in the target. However, it can also be seen that the general shape of the neutron production distribution curve has not changed much and that the neutron production peak is shifted by 1-2 mm only. Thus, the presence of Hg does not have a significant influence on the position of the peak, and the position at which the analytical optimization was started could be determined with a reasonable accuracy.

Figure 6.2 shows the energy dissipation per unit volume for the U-Mo block of $3 \times 3 \times 5$ cm³ for two different cases. The first one is the energy dissipation along the axis of the target for volumes $1 \times 1 \times 1$ mm³ located on this axis. The maximum observable value is 7.24 GW/m³. The second is the power density distribution averaged over each of the plates, with a highest value of 0.68 GW/m³. From the comparison of these curves it can be seen that there is a difference of a factor of 10 at maximum, which is given by a large inhomogeneity of the power density in the direction perpendicular to the electron beam. In total, a power of Q = 9350 W was deposited in the geometry. The reader will recognize that the volumetric power density curve shown in Figure 6.2 is very similar in shape to the neutron production curves from Figure 6.1. Both the neutron production and power density are governed by development of the electromagnetic cascade in the target (see chapter 4).

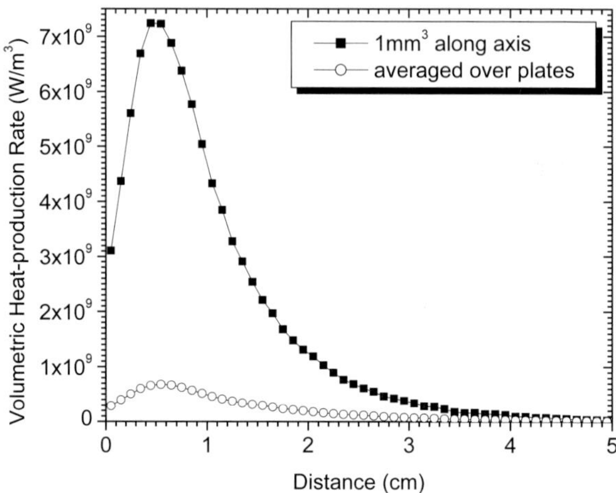

Figure 6.2. **Volumetric power density in U-Mo block of 3 x 3 x 5 cm³ a) along the axis, where the maximum value reaches 7.24 GW/m³, b) averaged over plates, with a maximum value of 0.68 GW/m³.**

To determine the thickness of each plate, analytical heat production/removal calculations were performed with the plate thickness as an independent variable, based on the MCNP-calculated power density value at given position in the U-Mo block. In this way, the maximum allowable thickness of each plate was obtained. The analytical approach is discussed below.

In order to calculate analytically the thickness of each plate separately, additional assumptions were made:
- the q value, which represents the heat generation rate in one half of the plate was calculated from the energy dissipation curve averaged over plates. However, there is a difference of a factor of 10 at maximum, compared to the curve for the energy dissipation along the axis of the target. This leads to an

underestimation of the maximum value in the plate, and therefore an additional correction is necessary. For this reason a modified energy dissipation value q* was used, calculated as $q^* = 5q$ (see Equations 6.3 and 6.4). In the case of using the energy dissipation along the axis of the target a large overestimation would result,

- the material properties are independent of temperature,
- the temperature distribution $T(x)$ in the plate is assumed to depend on the coordinate x only (x increases with increasing thickness of the plate),
- The heat transfer coefficient h was calculated from the Nusselt number $N_U = 3.76$, which is valid for two parallel plates and laminar flow ($h = 3.76\lambda_{Hg}/X = 62416$ Wm^{-2}K^{-1} for $\lambda_{Hg} = 8.3$ Wm^{-1}K^{-1} and $X = 0.5$ mm).

The channel temperature $T_{channel}$ and the wall temperature T_{wall} can be expressed as follows:

$$T_{channel} = T_{inlet} + \frac{q^*}{\Phi_m c_p} \; ; \qquad T_{wall} = T_{channel} + \frac{q^*}{hA_s} \qquad (6.3)$$

Then the central temperature of the plate T_{center} can be calculated by

$$T_{center} = \Delta T + T_{wall} = \frac{q^* D}{2\lambda A_s} + \frac{q^*}{hA_s} + \frac{q^*}{\Phi_m c_p} + T_{inlet} \qquad (6.4)$$

where D is the half thickness of the plate, λ is the U-Mo thermal conductivity (25 Wm^{-1}K^{-1}), A_s is the side area of the plate (9 cm^2), Φ_m is the mass flow rate of the coolant ($\Phi_m = v_{Hg}S\rho_{Hg} = 0.41$ kgs^{-1} for the coolant velocity $v_{Hg} = 1$ ms^{-1}, the channel cross section $S = 3 \times 10^{-5}$ m^2, and the coolant density $\rho_{Hg} = 13529$ kgm^{-3}), c_p is the specific heat of the coolant ($c_p = 138$ Jkg^{-1}K^{-1} for Hg), and T_{inlet} is the Hg inlet temperature (350 K). ΔT in Equation 6.4 represents the temperature drop over the thickness of the plate.

Equation 6.4 was used to obtain the thickness of each plate. However, due to approximations and uncertainties of the analytical calculations (estimated uniform power density in the plate, temperature independent material properties, estimated convective heat transfer coefficients etc.), the maximum temperatures in some of the plates exceeded 450 °C, as realized from later CFD simulations (see next section). Therefore an additional correction on the thickness of some plates was necessary and was carried out based on the results obtained with the CFD code FLUENT. Further, a Ta cladding was added to each of the U-Mo plates to provide the required containment of radioactive material, and to avoid the direct contact of the U-Mo alloy with the Hg coolant. Finally, a modification of the shape of inlet and outlet channels was performed in order to minimize the amount of Hg near the target. In this way, the final target design was obtained.

6.2   Computational fluid dynamics simulations with the FLUENT code

In order to simulate more realistically the heat transfer to the flowing mercury, the CFD code FLUENT [Flu01] has been employed. In CFD, the discretized conservation equations for mass (continuity equation), momentum (Navier Stokes equations) and thermal energy [Bir02] are solved numerically to model the physical behavior of the flow and the convective heat transfer.

Generally, the first aspect of a CFD treatment is a domain discretization, i.e. the volume being analyzed is subdivided into numerous smaller parts, the so-called cells. These cells constitute a mesh, the quality of which has a direct influence on the accuracy of the CFD simulations. The above mentioned conservation equations are discretized and solved for each single cell. There are various ways to discretize the equations, e.g. through finite differences, finite elements or finite volumes approaches [Zie00]. Fluent applies the latter. Within a cell the quantities are assumed to be constant. Thus, the choice of the size of the cells should reflect the expected spatial variations in the investigated quantities.

The following sections introduce the main features of the FLUENT code, the GAMBIT preprocessor [Flu01], which has been used in the initial step of the CFD investigation, as well as the main requirements for obtaining accurate CFD results.

6.2.1   FLUENT code

FLUENT is a proprietary, vendor supported, general purpose fluid flow modeling tool, designed to be able to model single-phase and multiphase flows, both laminar and turbulent, compressible or incompressible, with or without chemical reactions and heat and mass transfer, in two-dimensional or three-dimensional complex geometries [Flu01]. For all types of flows, FLUENT solves conservation equations for mass and momentum. In addition, also the equation for energy conservation is solved, if the problem involves heat transfer or compressibility.

Fluent provides great mesh flexibility, with unstructured meshes available to be generated in complex geometries. FLUENT also allows users to adapt (refine or coarsen) the grid based on the flow solution.

The capabilities of FLUENT that are most relevant to the present research are:
- complex 2D or 3D geometries
- laminar, and turbulent flows, with a number of different turbulence models available
- convective heat transfer, either natural or forced convection
- coupled conduction/convective heat transfer

6.2.2 GAMBIT preprocessor

The GAMBIT software [Flu01], available in the FLUENT package, was used in the project for geometry modeling, mesh generation, and boundary condition specification. GAMBIT has a single interface for geometry creation and meshing. Advanced tools are available to edit and replay building sessions for parametric studies. GAMBIT can import geometries from virtually any CAD/CAE software. Using the geometry tools, the fluid domain can be extracted from the imported geometry and a further decomposition can be performed.

6.2.3 Accuracy of the computational fluid dynamics simulations

Generally, the accuracy of the CFD simulations mainly depends on:
- the accuracy and suitability of the physical models used,
- the numerical accuracy of the simulations.

For laminar, single-phase flows the simulations are based on the direct solution of the conservation equations only. The accuracy of such simulations is limited by numerical accuracy only, i.e. by grid cell and time step size and the accuracy of the applied discretization schemes.

For more complex flows, such as multiphase and/or turbulent flows, semi-empirical model equations are added to the conservation equations, mainly to model phenomena that take place on such small length scales that they cannot be resolved by the numerical mesh. For such types of flow, inadequacies of the applied models add significantly to the numerical inaccuracy. Moreover, little general knowledge is available on the accuracy of the various models applied for many realistic flow configurations. In such cases, internal sensitivity analysis and/or experimental validation should provide indications as to the accuracy of the simulations.

6.2.3.1 Accuracy and suitability of applied flow models

Concerning the first issue, mainly the temperature-dependent physical properties of all target materials involved and the turbulence modeling played an important role in this project. There is a sharp difference between laminar and turbulent flow conditions. In laminar flow, fluid motion is highly ordered in streamlines in the direction of fluid motion. In contrast, fluid motion in the turbulent flow is highly irregular and is characterized by velocity fluctuations. These fluctuations enhance the transfer of momentum and energy, and thus increase the convective heat transfer rate. This is the reason why turbulent flow is superior to laminar flow, from a heat removal point of view.

As a criterion to determine whether a flow is in the laminar or in the turbulent regime, a dimensionless number, the so-called Reynolds number Re, is introduced

$$\mathrm{Re} = \frac{\rho v L_{char}}{\mu} \tag{6.5}$$

Here, ρ is the fluid density, v is the mean fluid velocity, L_{char} is the characteristic length of the domain, and μ is the dynamic viscosity of the fluid. The Reynolds number represents the ratio of inertial forces to viscous forces. If the behavior of a fluid depends mainly on its viscosity, the flow is laminar (low Re). However, if the momentum of the fluid has a much stronger effect on the flow behavior than the viscosity, the flow is turbulent (high Re). The flow is then dominated by inertial forces, producing random eddies, vortices and other flow fluctuations. Such a flow is highly unsteady. For intermediate values of Re the flow is transitional, i.e. partially laminar and partially turbulent. The transition between the laminar and turbulent flow starts to occur when the critical Re (Re_{crit}) is reached. The value for Re_{crit} depends on the exact flow configuration and is usually determined experimentally. To illustrate the wide range of Re_{crit} values, for a circular pipe it is 2,500 (internal flow), while for a flat plate it is 500,000 (external flow).

In the transition regime the flow is neither fully laminar nor fully turbulent. As a consequence, it is difficult to predict the fluid behavior, and therefore none of the flow models can provide a very accurate solution. To resolve this problem, one can apply the laminar model, as well as different turbulence models, to find the error margin of the model predictions. Generally, the computational stability of the CFD models is built on a numerical homogenization within the mesh cells. As viscosity is a physical mechanism for homogenizing flow variations, a problem can occur to distinguish between numerical and physical smoothing. This is very important when Re_{crit} problems are confronted, because they require a very accurate estimate of viscous stresses to obtain reliable results.

6.2.3.2 Numerical accuracy of computational fluid dynamics simulations

A second dominant source of error in CFD simulations for multidimensional domains is the numerical accuracy. Numerical inaccuracies arise from truncation errors that are a consequence of representing the flow equations in discrete form, both in time and in space. The spatial discretization error is determined by the mesh quality, therefore special attention needs to be paid when the mesh is built. More specifically, the CFD user should be aware of the fact that the error level is inversely related to the mesh resolution. Therefore, in order to minimize this error, one should perform an incremental refinement of the mesh to find a mesh-independent solution. The behavior of the quantities of interest should be examined for various meshes to obtain realistic CFD results.

In addition to the mesh quality and refinement, the numerical accuracy is also determined by the discretization scheme. It is the scheme used to interpolate flow values and fluxes between grid points. In general, lower order schemes are more stable but less accurate, introducing a large, unphysical, so-called numerical diffusion. The numerical diffusion is minimized when the flow is aligned with the mesh, and it is most observable for the cases in which the convection is dominant, i.e. when physical diffusion is negligible.

The first point is most important when choosing the mesh type. Clearly, a triangular or tetrahedral mesh can never fulfil the mentioned condition. A quadrilateral or hexahedral mesh, on the other hand, can be aligned with the flow only for simple flow situations. For complex flows, however, it is much more difficult to reach the mesh alignment with the flow even for the latter mesh type, and thus the mesh independence study becomes very significant.

6.3   Further target optimization using coupled FLUENT and MCNP

As mentioned earlier, accurate heat transfer simulations were required to reveal the thermal behavior of a new design. In addition, such simulations helped in further target optimization owing to more realistic results, compared to the analytical calculations.

6.3.1   MCNP heat source and interpolation to Fluent

The use of the FLUENT code for obtaining accurate results has implied the use of a realistic heat source. The heat source q''' represents a spatial distribution of the heat production in the target material, including the coolant, and was computed by the MCNP4C3 code. In all optimization cases, the usual 1-cm diameter electron beam with energy range between 70 and 140 MeV was applied. Naturally, the power density distribution in the target is case-dependent, and therefore had to be calculated for each case separately. This was done in MCNP by splitting the target geometry into small volumes, and by calculating the heat source value for each of these volumes. When the block-shaped new target is split into several U-Mo plates, thin plates are required near the location of maximum power dissipation, while thicker plates are allowed in the parts of the target in which the power density becomes low (see Figure 6.2).

As a second step, to provide FLUENT with the correct heat source, an interpolation was performed between the MCNP and the FLUENT mesh. This step was required as the number (and thus the size too) of the MCNP volumes, for which the heat source was calculated, did not correspond to the number of cells comprising the FLUENT mesh. In order to perform the interpolation with a high accuracy, a decision has been made to use a block-shaped mesh for both MCNP and FLUENT simulations.

FLUENT provides users with the opportunity to define arbitrary volumetric sources of heat. This particular feature was used to specify the power density distribution in the target previously calculated by MCNP. In order to do so, special **User-Defined Functions (UDFs)** [Flu01] were written and implemented in the FLUENT simulation process (see Figure 6.3). The first UDF calculates the heat source value for a given cell, and stores the value at a **User-Defined Memory (UDM)** array [Flu01]; the second reads the values from the UDM array and transfers to FLUENT for further processing. The volume-weighted technique was chosen for the interpolation to conserve the total deposited energy in the target.

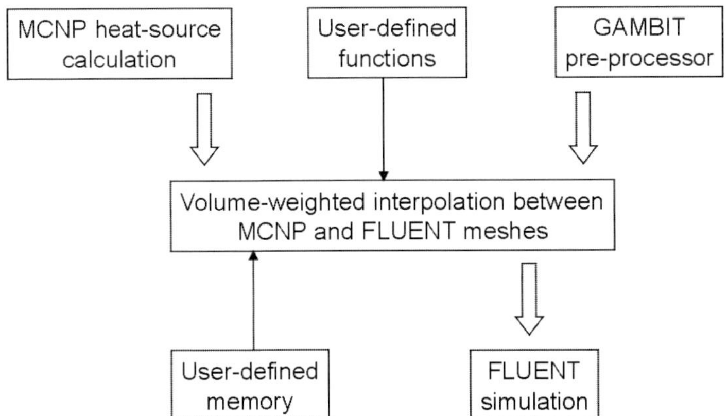

Figure 6.3. The heat source volume-weighted interpolation between MCNP and FLUENT.

6.3.2 *Assumptions applied in the FLUENT optimization*

In all CFD simulations, the following conditions and assumptions were involved, unless stated otherwise:

- Hg with a constant mass flow rate of 2.25 kg/s was flowing through the geometry. This corresponds to the nominal flow rate of 10 l/min that may be realized by the electromagnetic pump, presently used at GELINA. This flow rate has been ensured by a mass-flow boundary condition defined on the inlet.

- The inlet Hg temperature was kept constant at 350 K, which corresponds to the current situation at GELINA.

- Adiabatic external walls were assumed, i.e. no heat transfer was allowed through the domain boundaries to the target surroundings. This assumption implies a slight overprediction of the maximum temperature by the CFD calculations, as, in practice, the heat transfer through the domain walls is not zero. However, it can be considered as negligible. The consequence of this condition is that only the Hg coolant can remove the heat from the target.

- The FLUENT solver calculated heat transfer from the solution in the cells adjacent to a wall between solid and liquid regions. For laminar flow simulations, this implies a straightforward heat transfer calculation without additional model assumptions. For the turbulence simulations, depending on the applied turbulence model, additional modeling is needed to model heat transfer through the laminar boundary layer adjacent to the wall. This is discussed in more detail in section 6.5.1.

- All results presented here were obtained by using second-order spatial discretization schemes for energy, momentum, and turbulence quantities (for turbulence modeling).

6.4   Initial target design – results and discussion

As a first step towards the new GELINA target, the U-Mo block was split into ten 3×3 cm^2 plates of varying thickness such that the dissipated power per plate was below 1 kW (see Figure 6.4). This segmentation, however, was not an optimal one, as explained earlier. Nevertheless, this design was tested from the neutronics point of view, and the results were very promising (see chapter 5). Therefore the most important CFD results for this design are presented in the following text.

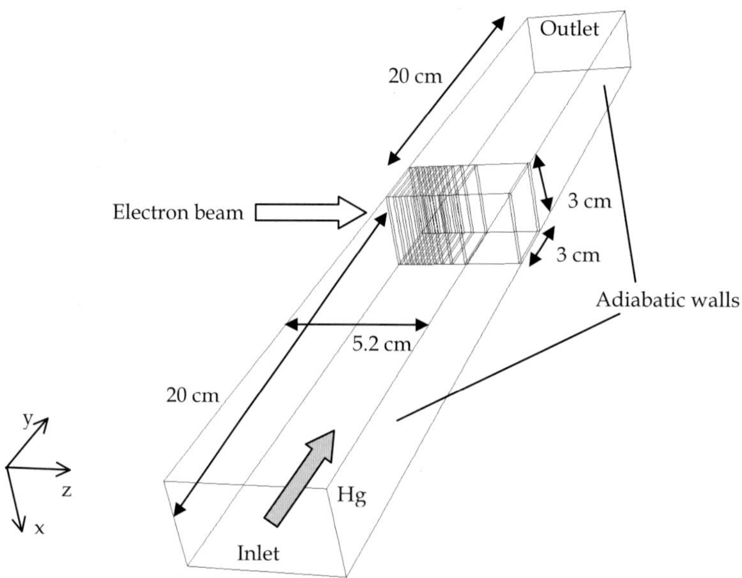

Figure 6.4.   The initial target design. The domain consists of ten 3 x 3 cm² U-Mo plates of various thicknesses, with eleven 1-mm Hg cooling channels, and two 20-cm Hg blocks placed above and below the target.

The total length of the initial target design is 5.2 cm, and it starts with a Hg cooling channel. The first and last cooling channels border with the adiabatic walls. As a starting point, two 20-cm-Hg blocks were placed above and below the target to obtain relevant information about the flow behavior in the domain (see Figure 6.4). The total number of FLUENT cells used reached 750,000, including 50,000 cells in the target itself (U–Mo plates plus 1-mm Hg channels). It has to be stressed that only the laminar flow model was used here for the CFD simulations, and no additional mesh refinement was carried out to improve the numerical accuracy. This technique was employed only for the final design, as will be described later. In this case, the Reynolds number for the inlet equals 37,000. The Reynolds number for a single cooling channel is about 9,600, which indicates that the flow is not fully turbulent in the space between the target plates.

Figure 6.5 depicts the temperature distribution of the upper part of the investigated domain, on the midplane in the direction of the flow.

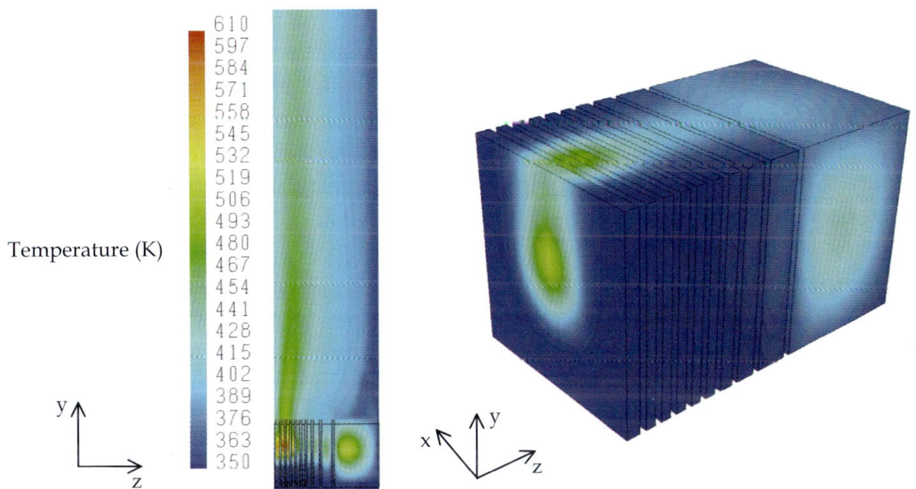

Figure 6.5. Temperature distribution on the midplane in the direction of the flow (left). Hg flows from the bottom to the top of the domain. Two 20-cm Hg blocks are located above and below the target (only upper Hg block is shown). 3D temperature visualization of the initial design (right) reveals large temperature variations in both axial and radial directions.

The lower part is not shown in Figure 6.5, because there the temperature is constant and identical to the Hg inlet temperature (350 K). The highest temperature reached is 606 K (333 °C), occurring in the second plate of the geometry. This value is far below the chosen temperature limit (723 K). It can also be observed in Figure 6.5 that the maximum temperature values in each plate lies close to the target axis. Naturally, the power density is highest along this axis as a consequence of the development of the electromagnetic cascade (see chapter 4). A small shift in the direction of the flow is apparent, which is caused by the heat convection process to the coolant flowing between the plates, and consequently higher coolant temperatures in the upper parts of the 1-mm channels. As can be concluded from the earlier results, most of the heat is dissipated in the first part of the target (see Figure 6.2). Due to the high temperatures in the first four U-Mo plates, a wake of hot Hg is created above the target, which can be clearly seen in Figure 6.5. This wake slowly disappears with increasing distance from the target as a consequence of the thermal equalization between the Hg streams of different temperature.

A detailed view of the temperature distributions in the target is shown in Figure 6.6 (see p. 92). The midplane cuts in the direction of the flow and perpendicular to the flow are presented. The latter case manifests the temperature symmetry along the target axis, which is given by the shape of the power density distribution (see Figure 6.7, p. 92), and the symmetrical boundary condition for the flow around z axis. The total power present in the target was 9,380 W. No heat source was considered in the Hg blocks above and below the target.

The temperature distribution along the central axis of the target is shown in Figure 6.8 (see p. 93). The maximum temperatures in the first three plates are similar, and reach values up to 590 K. The maximum temperature reached at the central axis is a bit lower than the absolute maximum in the domain (590 K versus 606 K). This is caused by the slight downstream shift of the temperature maximum in the flow direction. The maximum temperature at the axis decreases with increasing distance from the electron source impact area, with the minimum value below 450 K in plates 6 and 7.

Based on these results, it was concluded that there was a potential to further optimize the target design, especially in the central part where the temperatures were much lower than the allowed maximum.

The velocity profiles for the Hg flowing through the cooling channels, as presented in Figure 6.8, reveal that no significant differences can be expected between the cooling channels. The maximum velocity varies in the range 0.5-0.6 m/s in the channels. This slight difference is caused by different thicknesses of U-Mo plates, mainly by the plate with a thickness of 2.15 cm. This plate forms a large resistance to the flow, which results in higher Hg velocities in the channels adjacent to this plate.

The velocity vectors on the midplane in the flow direction are depicted in Figure 6.9 (p. 93). Due to the size of the last U-Mo plate, a strong recirculation zone is created before and behind the plate. This plate forms a large geometrical resistance to the flow. Before the plate a stagnation pressure builds up, which is of the order of magnitude of the dynamic pressure $1/2\rho_{Hg}v_{Hg}^2$. At the edge of the plate this quantity is transformed into the extra kinetic energy. Behind the plate a partial conversion back to pressure occurs, while the rest of the energy is consumed by eddies in recirculation zone. Likewise, all other plates also exhibit small recirculation zones behind the plates, thus they also cause some loss of the energy owing to the blocking of the flow. This energy loss influences the final pressure drop of the design. However, in the case of the initial design the Fluent flow simulations showed this pressure drop to be only 3 kPa.

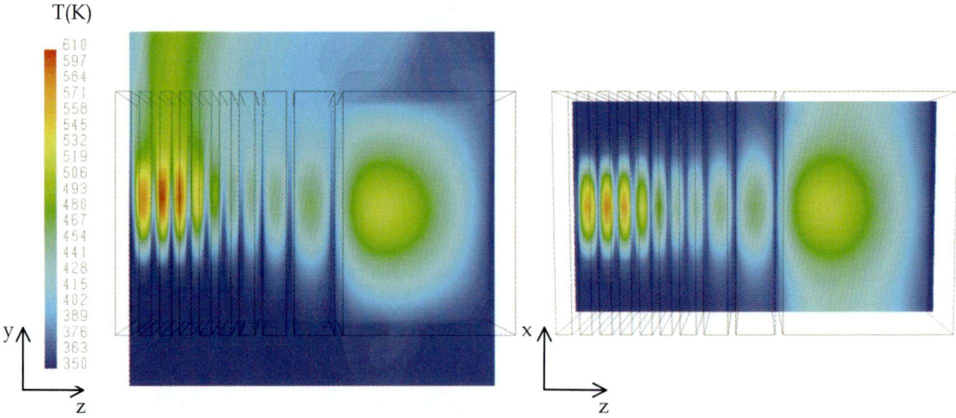

Figure 6.6. **Temperature distributions on the midplanes in the direction of the flow (left) and perpendicular to the flow (right).**

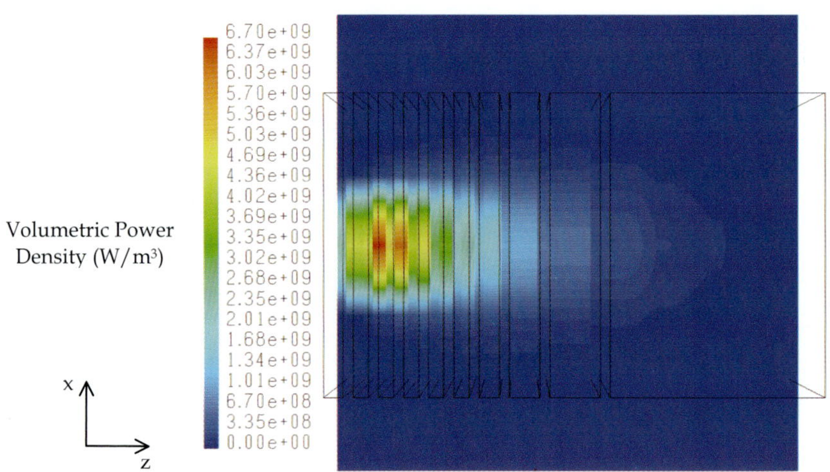

Volumetric Power Density (W/m³)

Figure 6.7. **Visualization of the FLUENT heat source, as calculated by the MCNP4C3. The maximum value reaches 6.7 GW/m³.**

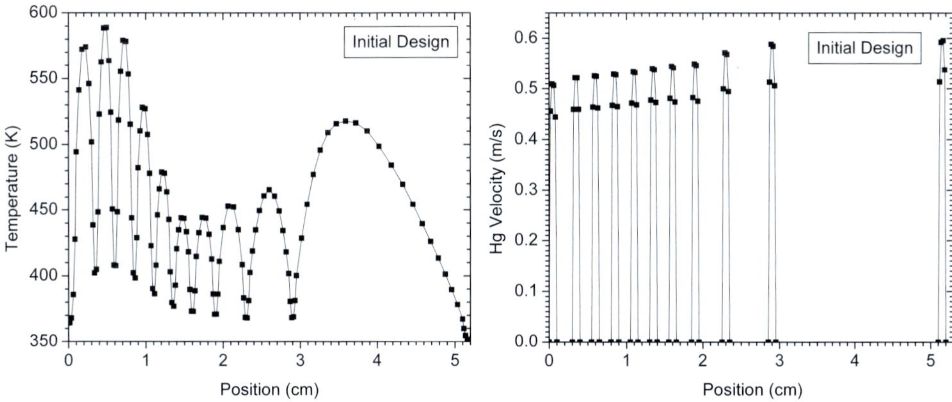

Figure 6.8. Temperature (left) and Hg coolant-velocity (right) distributions along the z axis. The minimum temperatures are reached in the 1-mm Hg cooling channels. No strong variation of maximum velocities is observed.

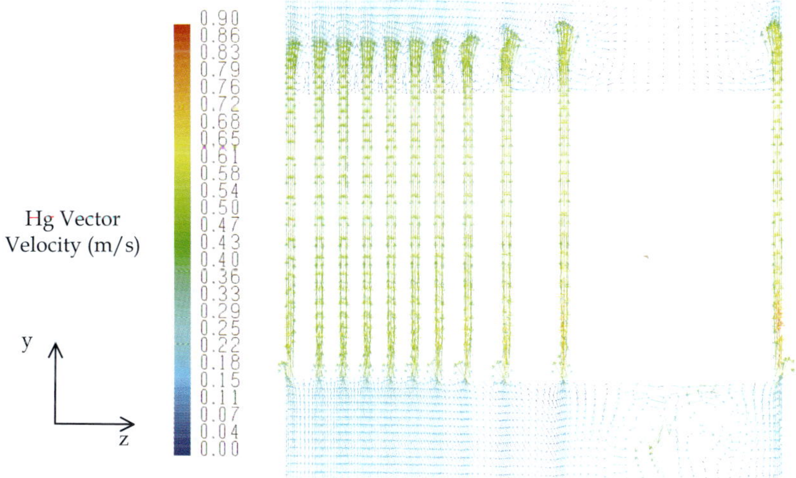

Figure 6.9. Velocity vectors on the midplanes in the direction of the flow. Recirculation zones are observed in the flow, especially before and behind the last U-Mo plate.

6.5 Final target design

In this section we present CFD flow and heat transfer simulation results for the final design, which was obtained by using the optimization procedure described earlier in this chapter. The design consists of seven 3 x 3 cm² U-Mo plates with thicknesses of 0.15 cm, 2 x 0.2 cm, 0.3 cm, 0.45 cm, 0.9 cm, and 2.1 cm (see Figure 6.10). Ta cladding with a thickness of 0.2 mm coats each U-Mo plate. This coating creates containment of radioactive material and serves to prevent U-Mo from being in direct contact with Hg. In this way, diffusion of the radioactive material from the plates will be minimized. Seven 1-mm Hg cooling channels were positioned between the plates. The length of the target, excluding the inlet and outlet channels, is 5.28 cm. The geometry begins with a U-Mo plate, which is cooled from one side only. In contrast with the initial design, the thickness of the inlet and outlet Hg channels was minimized to 0.5 cm above and below the target (see Figure 6.10), with a length of 5 cm. The reduction of the coolant volume has a consequence in higher FOM values, in which especially the contribution of the improved RFs plays an important role.

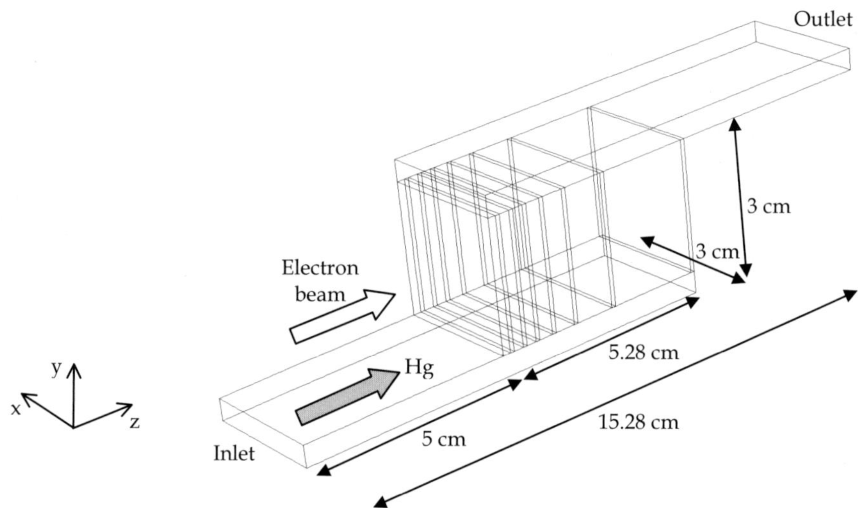

Figure 6.10. The final target design. The domain consists of seven 3 x 3 cm² U-Mo plates of various thicknesses, seven 1-mm Hg cooling channels, and two 0.5-cm Hg blocks placed above and below the target. 0.2-mm Ta layers coat the U-Mo plates.

6.5.1   CFD model settings

As discussed earlier for the initial design (see section 6.4), the values of the Reynolds numbers for the Hg flow in the inlet and outlet channels and through the cooling channels are such that it is not unambiguously clear whether the flow is laminar, fully turbulent, or in a transitional regime. The actual flow regime has a very strong impact on the heat transfer calculations. Heat transfer rates are generally lower for laminar flow than for turbulent flow. The ambiguity of the nature of the flow poses a problem for CFD flow simulations with a code like Fluent, where the user has to specify whether the flow is laminar or turbulent. In the latter case, an approximate turbulence model must be applied, of which there are various available in Fluent. All turbulence models must be considered as rather crude approximations of physical reality. Each has its particular strengths and weaknesses, and its most suitable range of applications.

For the complex geometry under study and the range of *Re* numbers occurring, however, no rigorously supported statements can be made as to the applicability and accuracy of the laminar or various turbulent flow models. Therefore, the engineering approach followed was to perform each flow and heat transfer simulation with both the laminar flow model and two different turbulent flow models: the so-called k-ω standard model, and the k-ε RNG model [Flu01]. In both models, in addition to the conservation equations for mass, momentum and energy, two additional model equations are solved from which the characteristic velocity and length scale of turbulence fluctuations are estimated. From these, a so-called eddy or turbulent viscosity (μ_t) is calculated, which is added to the molecular viscosity in the momentum balance equations. It is expected that different results will be obtained for the laminar and the two turbulence models. The spread between the three is a rough indication of the uncertainty in the results.

Two additional issues relating to the turbulent flow simulations need to be discussed here:

The first is the inlet boundary condition for the so-called turbulence intensity, defined as the ratio of the root-mean-square of the velocity fluctuations to the mean flow velocity. This turbulence intensity is unknown, since it fully depends on the conditions of the flow upstream, i.e. outside the simulated domain. On the other hand, its value is convected into the computational domain and may have a strong influence on the simulation results. For a fully developed turbulent flow in a long pipe, the turbulence intensity is ~4%. In all our turbulence simulations, we assumed the flow to be underdeveloped at the inlet, with a turbulence intensity of 1% at the inlet. It will be shown later that no significant change of the temperature profile in the target is caused by using higher turbulence intensities (10%, see Figure 6.11 later in the text).

The second issue is the treatment of the near-wall region. The near-wall region can be divided into three main regions, namely the viscous (laminar) region, the mixing region, and the turbulent region. In many practical flow situations, like the one under study here, it is not feasible to use a computational grid that is fine enough to fully resolve the near wall region. Therefore, as an alternative, so-called wall functions [Flu01] can be used to bridge the gap between the turbulent core of the flow and the solid walls. This approach was applied in the present study together with the k-ε model. The proper use of these wall functions imposes restrictions with respect to the distance between solid walls and the first adjacent grid point. This is further discussed below. The k-ω model, however, does not use wall functions, but instead it applies a special near-wall modeling technique [Flu01]. The k-ω model was designed to be applied throughout the near-wall region, provided that the resolution of the near-wall mesh is sufficient.

6.5.2 *Accuracy and grid dependence of the results*

Not only the proper choice of physical models, but also the choice of the mesh size has a significant impact on the simulation accuracy. To obtain mesh-independent results, various meshes were investigated. The mesh configurations are described in Table 6.1. Since the temperature behavior in the target is of the highest interest, this quantity has been selected for judging the mesh independence of the FLUENT results. This was achieved by comparing the distributions along the central axis of the target in the direction of the electron beam.

Mesh number	1	2	3	4
Number of mesh cells	254,000	881,000	2,511,000	6,274,000
Cell resolution in 1-mm channel (cells)	5	10	15	20
Cell resolution in 5-mm channel (cells)	15	20	30	40

Table 6.1. **Mesh configurations used to investigate the mesh independence of the FLUENT results. Cell resolution number represents the number of cells in the direction perpendicular to the main direction of flow.**

In Figure 6.11 a comparison is given of the temperature distributions along the central axis of the target. The Reynolds number in the inlet equals 85,000, and in a single cooling channel it is approximately 18,000. For meshes 1, 2 and 3 FLUENT simulations were performed using the laminar and two turbulence models. For Mesh 4, only the laminar model was used. A comparison of the results for the laminar model on the 4 different meshes reveals that Mesh 1 provides only a rough estimate of the temperature distributions. This can be seen from the large temperature differences when compared to other meshes. The distributions for all other meshes are very similar. It has been observed that Mesh 4 did not bring any additional correction to the distribution shape. Therefore it was concluded that Mesh 3 provides a sufficiently fine discretization for the laminar model.

For the two turbulence models the temperature differences between the various meshes are lower, and are always below 50 °C when Mesh 2 and Mesh 3 are compared. It is also observed that the differences between these two meshes are lower for the k-ε model, as observed in the first three plates. Obviously, Mesh 3 represents a good choice for discretizing the target domain.

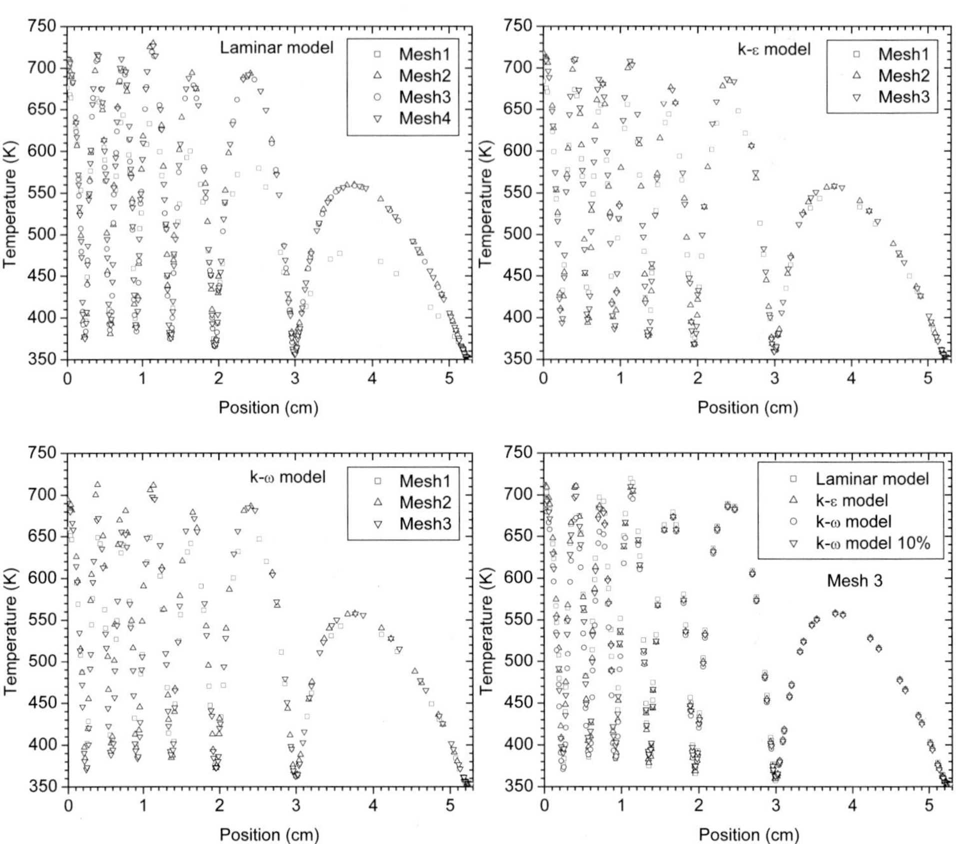

Figure 6.11. Temperature profiles along the axis of the target in the direction of the electron beam. Mesh 3 was found out to provide a reasonable accuracy in the simulations.

Mesh 3 comparisons in Figure 6.11 show only small discrepancies between the values estimated by different models, in which the difference never exceeds 40 °C. Based on these results, the temperature error has been estimated to be not more than 50 °C.

The mesh quality near the wall is of high importance and the accurate prediction of wall-bounded turbulent flows is achieved only if a proper representation of the flow is used in the near-wall region. The FLUENT user has an opportunity to judge the quality of the mesh by visualizing the distance from the wall to the wall-adjacent cells. This distance is usually given in wall units y+ or y* [Flu01]. The requirements on the y+ values were fulfilled for both turbulence models used with Mesh 3, which additionally approves the choice of this mesh for obtaining the final results.

6.5.3   Results and discussion

In the following text the most important results are presented as obtained by applying Mesh 3 in FLUENT simulations for the final design geometry. The 3D variation of the temperature in the target is shown in Figure 6.12. This result was obtained by using the laminar model. It is evident that the highest temperatures are reached close to the central axis of the target. In this case the maximum temperature is 723 K (450 °C).

The target heat source, as calculated by MCNP for the final design geometry, is shown in Figure 6.13. The highest value of the distribution is 6.78 GW/m³, which is similar to the maximum value for the initial design. No heat source has been implemented for the inlet and outlet Hg channels. The contribution of the heat deposited in these parts of the target is negligible, corresponding to about 100 W. The total power deposited in the target equalled 9,440 W.

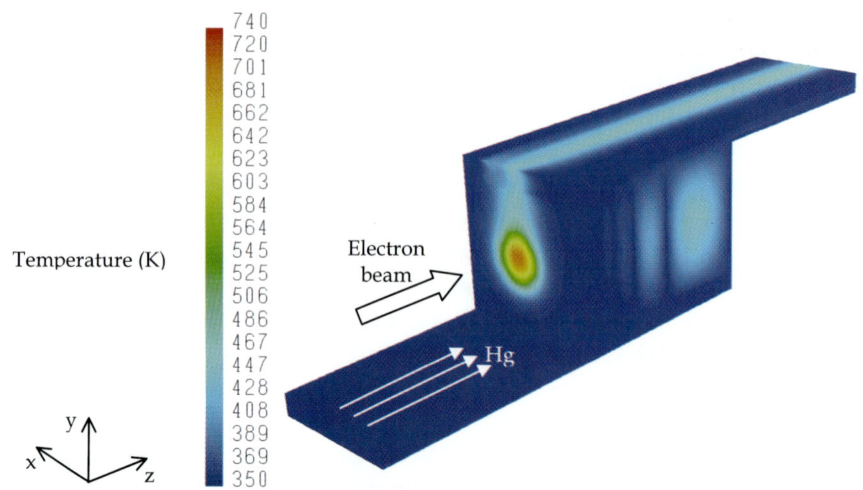

Figure 6.12. 3D temperature visualization of the final target design. The laminar model was applied.

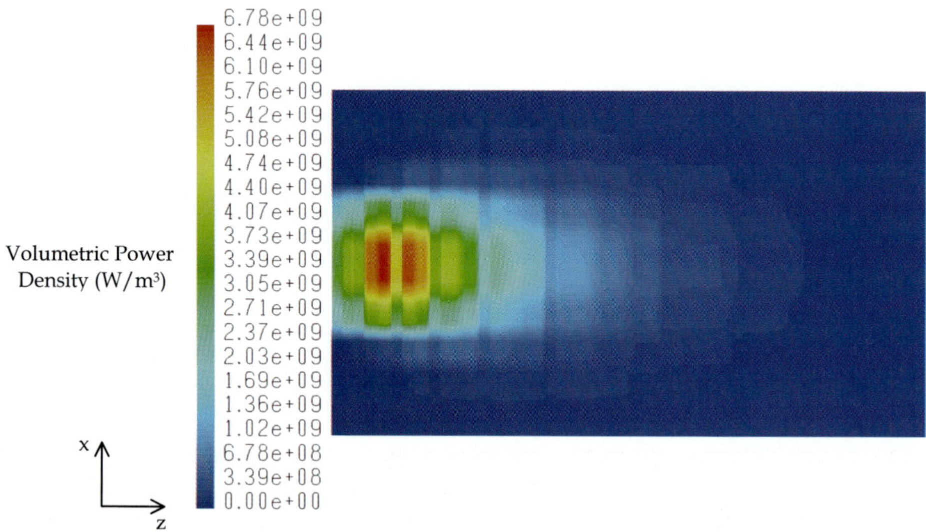

Figure 6.13. Visualization of the FLUENT heat source, as calculated by the MCNP4C3.

Figure 6.14 depicts the temperature distributions on the midplanes in the direction of the flow for the laminar and two turbulence models used.

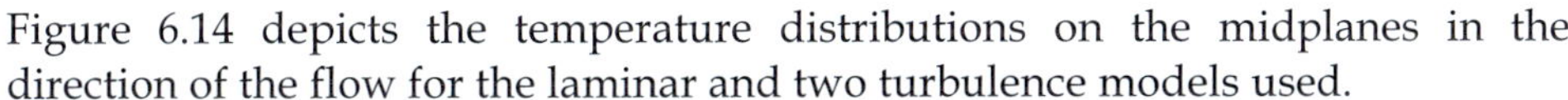

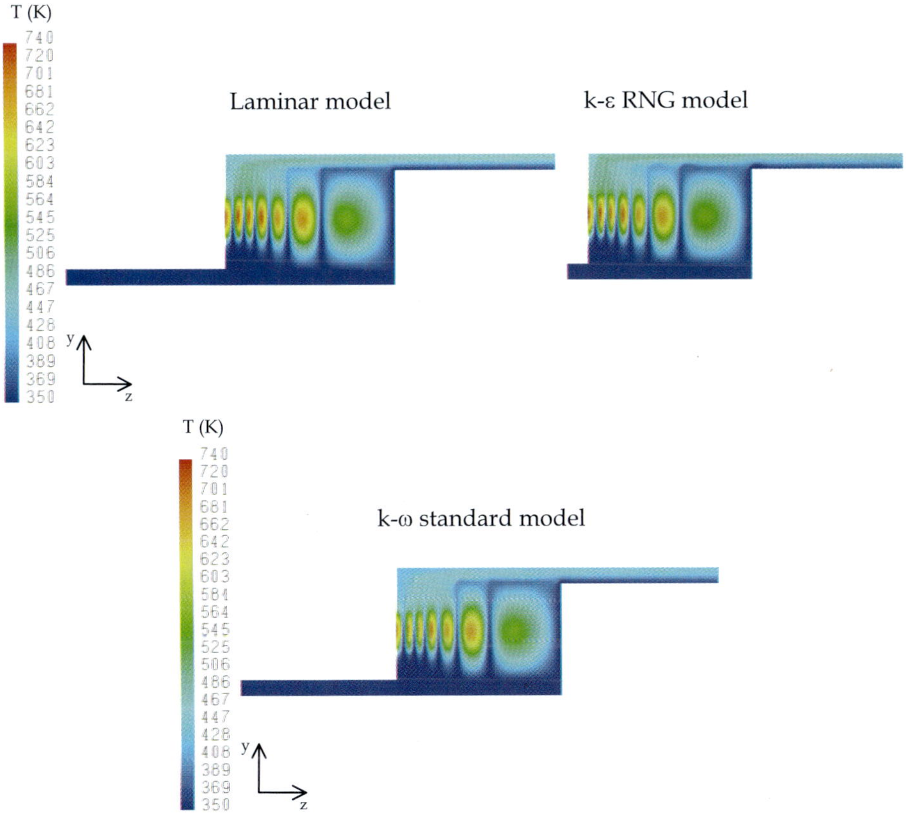

Figure 6.14. Temperature distribution on the midplane in the direction of the flow. The laminar model (upper left) was used, as well as the turbulence models k-ε (upper right) RNG and k-ω standard (lower).

The highest temperature is always reached in the fourth plate of the target. In the case of the laminar model, the highest temperature matches the chosen temperature limit for the optimization (450 °C). In the cases of the k-ε and k-ω models, the highest temperatures are 450 °C and 427 °C, respectively. The maximum temperatures are the same for the laminar and k-ε models. Generally, due to the higher heat removal efficiency of turbulent flow, it is expected to reach lower temperatures when using the turbulence modeling. However, the flow regime is not fully turbulent in this case (see *Re* numbers in section 6.5.2), which might explain the agreement between the maximum temperature values estimated by the laminar and the k-ε model.

Naturally, the power density is highest close to the target axis, which is caused by the shape of the power density distribution in the target (see Figure 6.13). Only a minor temperature shift from the central axis in the direction of the flow is shown in Figure 6.14, which is caused by the heat convection process.

It is also observed in Figure 6.14 that the major contribution to the temperature increase of the coolant comes from the first four plates. However, the Hg temperature increase between inlet and outlet is only about 30 K. It is also obvious that the maximum temperature in the last U-Mo plate, which is about 550 K, is far below the maximum allowable value. Nevertheless, the presence of this plate is important from the neutronics point of view, as its elimination would cause a decrease of the neutron flux at 90° by 20%. From this 20%, 10% is due to neutrons created in this last plate; the other 10% comes from scattering events.

Figure 6.15 (see p. 104) shows the temperature distributions on the midplane perpendicular to the flow direction. The heat dissipation is centralized and is strictly symmetrical with respect to the axis of the geometry. Only minor temperature discrepancies can be found among the FLUENT results for different flow models. Similar distributions, but at different planes perpendicular to the flow, are shown in Figure 6.16 (p. 104) to present the temperature variation in the radial direction. The results were obtained from the simulation using the laminar model. Four temperature distributions are shown in Figure 6.16. The plane y = 1.52 cm is the midplane of the target, with the total target thickness being 3.04 cm, excluding the inlet and outlet channels.

The velocity vectors are depicted in Figure 6.17 (see p. 105). Only slight differences are observed between the different flow treatments. The maximum velocity vector values are 2.07 m/s, 2.05 m/s, and 2.26 m/s for the laminar model, the k-ε model, and the k-ω model (not shown in Figure 6.17), respectively.

The nominal mercury flow rate, as provided by the present electromagnetic pump is 10 l/min. Consequently, all FLUENT simulations presented in this chapter were performed using this flow rate. However, it is also important to predict the behavior of the target from the thermal point of view during situations, in which the amount of the coolant could decrease below the nominal value. In order to assess this behavior, FLUENT simulations were performed for various Hg flow rates. Figure 6.18 (p. 106) depicts the results obtained using the laminar modeling for the flow, and Mesh 4 (see Table 6.1). As expected, the temperature would increase with a decreasing Hg flow rate through the target. It is interesting to find out that even a flow rate of 4 l/min would keep the maximum target temperature below the first phase change of U, taking into account an error margin of 50 °C estimated earlier in this chapter. Nevertheless, the operation with such a low flow rate is not recommended for safety reasons. Based on the results from Figure 6.18, an increase of the nominal flow rate would lead to a small decrease of the maximum temperature only. Therefore, no effort is necessary to increase the nominal flow rate value, as this modification would not lead to any significant improvements.

The total pressure drop over the target has been estimated from the FLUENT simulations (using the laminar model) to be about 16 kPa. For the k-ε and k-ω models, the pressure drops are 16 kPa and 23 kPa, respectively. The pressure drop values obtained using the turbulence model k-ω is higher, compared to the laminar and k-ε models. Generally, as the turbulent fluid flows through the domain it loses more energy than the laminar flow, owing to the fluctuations of the main flow quantities. This results in a higher pressure drop. An agreement between the pressure drops, as estimated by the laminar and k-ε models, can be explained by the fact that the fluid regime in the domain is neither fully laminar nor fully turbulent. Therefore none of the viscous models used can be considered as a superior over the other models. All pressure drop values for the final design are higher than the value for the initial design (3 kPa, laminar model). This is due to the different geometry configuration, resulting in a higher energy loss of the coolant passing through the final design target.

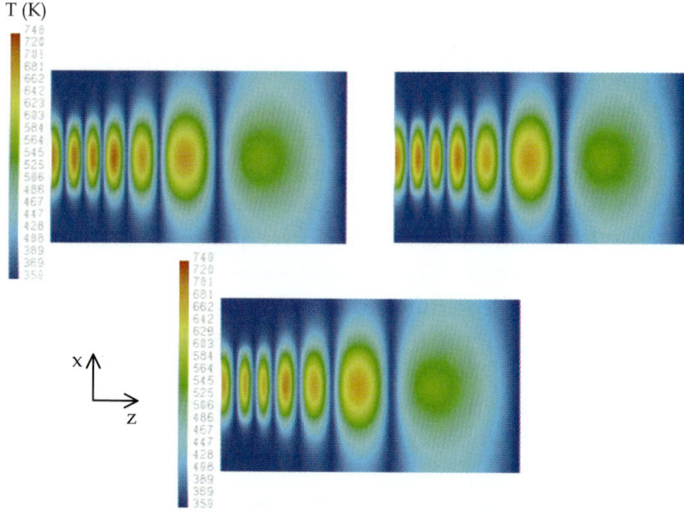

Figure 6.15. Temperature distribution for the laminar model (upper left), the k-ε model (upper right), and the k-ω model (lower) on the midplane perpendicular to the flow direction.

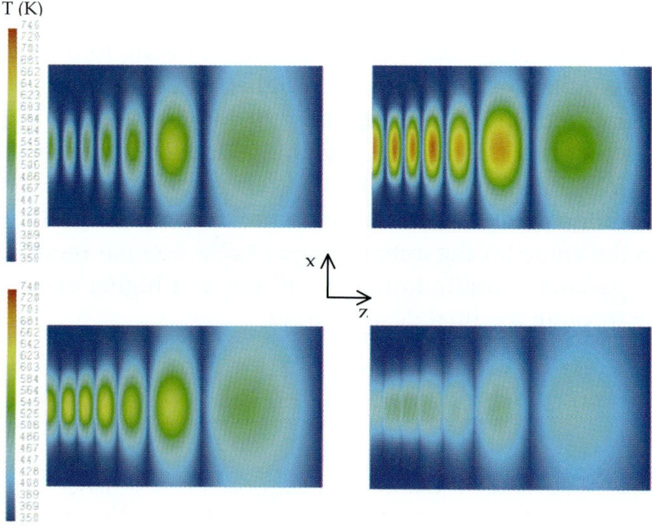

Figure 6.16. Temperature distribution for the laminar model on the planes perpendicular to the flow direction, at positions y = 1.02 cm (upper left), y = 1.52 cm (upper right), y = 2.02 cm (lower left), and y = 2.52 cm (lower right).

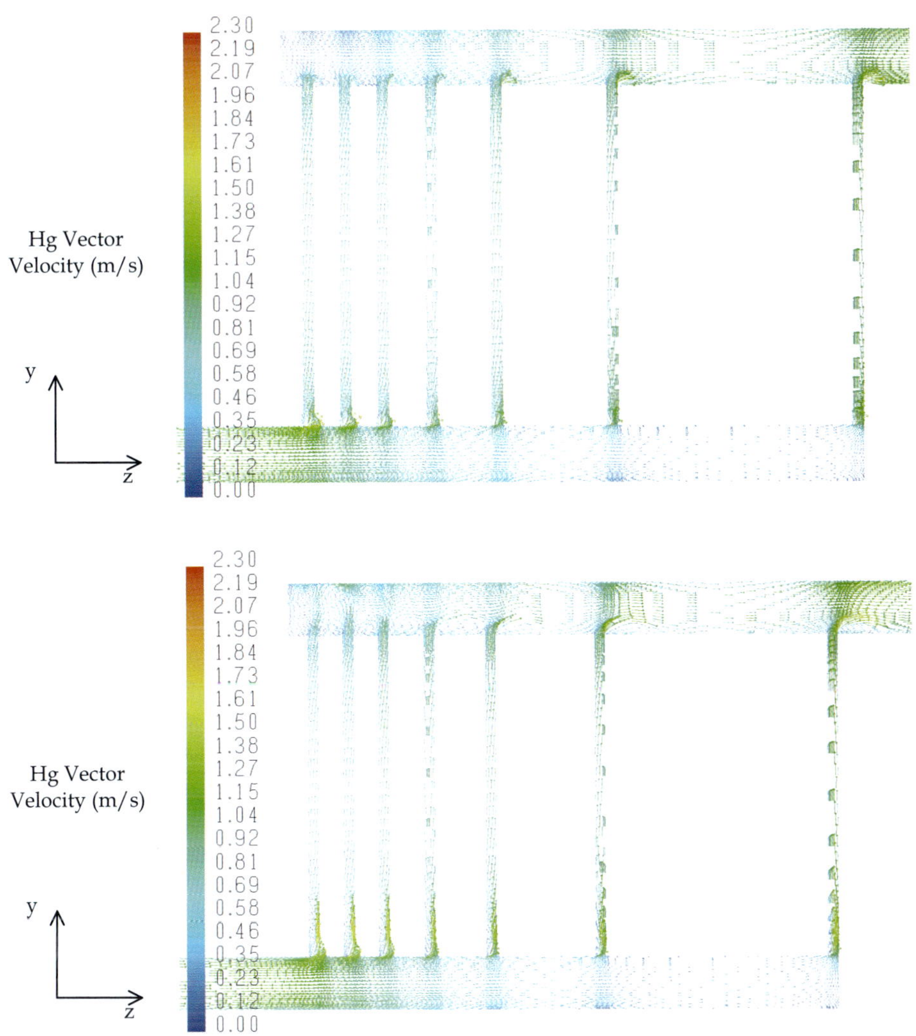

Figure 6.17. Velocity vectors on the midplanes in the direction of the flow for laminar model (upper) and the k-ε model (lower). No large-scale eddies are observed in the flow, and no significant difference is noticed between the laminar and turbulence model approaches. Small recirculation zones are observed for the k-ε model above the plates.

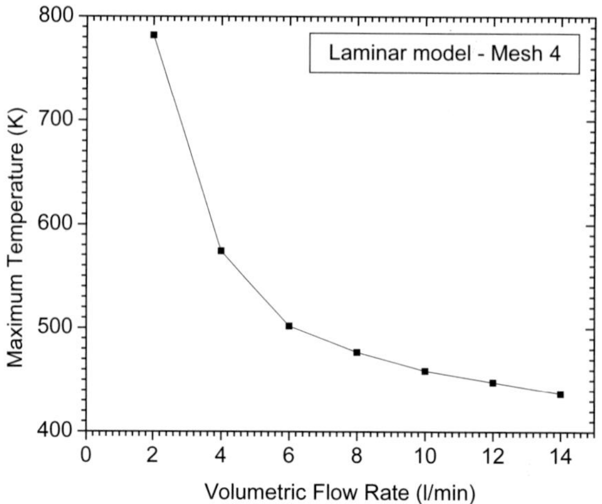

Figure 6.18. The variation of the maximum temperature in the final target design for different coolant flow rates. Results were obtained using the laminar flow model.

Finally, Table 6.2 summarizes the most relevant quantities discussed earlier for the final design using Mesh 3 and the nominal Hg flow rate (10 l/min). The coolant temperature increase between inlet and outlet is about 30 K in all cases. This mesh provides sufficiently fine discretization for all viscous models.

Final design	T_{max} (°C)	v_{max} (ms⁻¹)	Δp (kPa)
Laminar model	450	2.1	16
k-ε RNG model	450	2.1	16
k-ω Standard model	427	2.3	23

Table 6.2.   The quantities of interest for the final target design using Mesh 3 and a Hg flow rate of 10 l/min.

6.6   Conclusions

By proposing the new compact target design all project objectives are fulfilled. The new target was designed using a dedicated optimization procedure. In this way, the final target design was obtained, consisting of seven 3 x 3 cm^2 U-Mo plates with thicknesses of 0.15 cm, 2 x 0.2 cm, 0.3 cm, 0.45 cm, 0.9 cm, and 2.1 cm. Tantalum cladding with a thickness of 0.2 mm coats each U-Mo plate. The heat produced in the target is removed by mercury flowing in seven 1-mm cooling channels. The length of the target itself is 5.28 cm. The maximum temperature never exceeds the conservative limit of 450 °C in CFD simulations. However, since the fluid in the target is in the transition regime, it is not straightforward to predict the fluid behavior. Therefore we applied not only the laminar model, but also the k-ε and k-ω turbulence models to find the error margin of the model predictions. This error margin is estimated to be about 50 °C, thus the maximum temperature for given operation conditions will not exceed 500 °C. This temperature is still well below the first phase change of uranium (668 °C), which should be avoided in order to prevent target damage.

Chapter 7

Final conclusions

This thesis describes a design study with the purpose to further enhance the quality of the GELINA facility by proposing a new neutron producing target. It is demonstrated that there is a potential for an improved target to allow GELINA users to measure high-energy neutron cross sections with even higher accuracy. Therefore an effort was made to optimize the size, shape, and material composition of such a target design in view of the optimal neutron source characteristics, while providing an adequate solution for target cooling. In this chapter the final results are summarized to emphasize the most relevant conclusions of this study.

The MCNP simulation results show that a substantial reduction of the target volume, compared to the present GELINA target, will lead to improved resolution functions, while maintaining the overall neutron yield. The 4π neutron escape of the present rotary target, which is about 5.6% n/e can be fully maintained by a U-Mo cylinder with R = 1.5 cm, and H = 5 cm. Uranium is the best material from the neutronics point of view. U-Mo alloy with 10%wt. of molybdenum was chosen instead of pure metallic uranium for the rotary design to improve the mechanical properties of the target material. The presence of molybdenum has only a negligible impact on the neutronics properties of a compact U-Mo target, and therefore this alloy will be employed in a new target design. The 4π neutron yield is lower by 45%, if other material than uranium is used (tantalum, tungsten, mercury), using the cylindrical targets scaled by quantities specifying an electromagnetic cascade. The use of thorium would cause a 20% loss only.

Concerning the resolution functions, tantalum and tungsten are comparable with U-Mo. Both these materials could be reasonable substitutes for U-Mo to provide equivalent resolution functions. However, the use of one of these materials would lead to a large neutron loss. It was also realized that the resolution functions could not be improved above a neutron energy of 3 MeV. The improvement can be observed only in the tail of the resolution functions below the full width at one tenth of the maximum (FWTM), which is of very low importance.

The first phase change of uranium appears at 668 °C. At a nominal power level of 10 kW the maximum temperature of a single-piece U-Mo block would exceed this temperature, which would lead to target damage, caused by dimensional changes of the material. This was checked by analytical calculations. Therefore, a dedicated optimization procedure was used to reduce the maximum temperature in the target by its segmentation into right-angle plates perpendicular to the target axis. In this way, the final target design was proposed (see Figure 7.1). Mercury was chosen as a coolant to keep the temperature below a certain temperature. A temperature of 450° was chosen as a conservative limit.

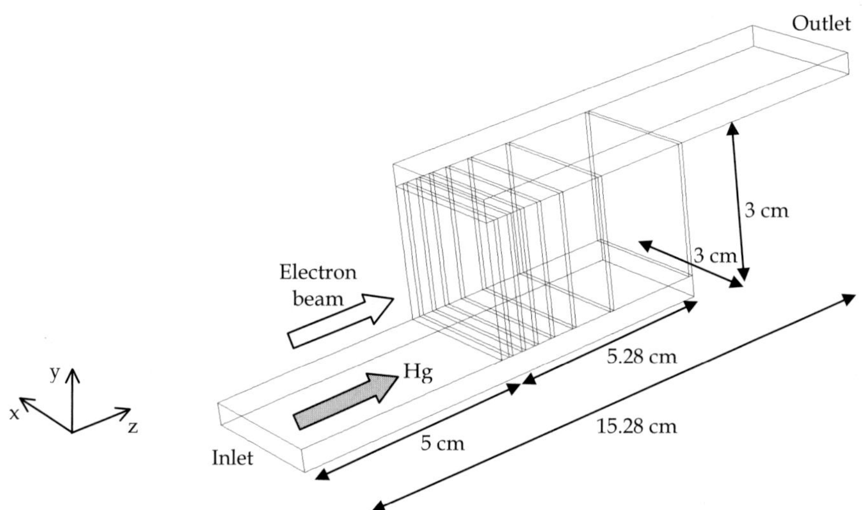

Figure 7.1. The new target design proposed.

The final design consists of seven 3 x 3 cm² U-Mo plates with thicknesses in the direction of the beam of 0.15 cm, 2 x 0.2 cm, 0.3 cm, 0.45 cm, 0.9 cm, and 2.1 cm. A tantalum cladding with a thickness of 0.2 mm coats each U-Mo plate. Seven 1-mm mercury channels located in between the plates cool the target. A channel thickness of 1 mm has been chosen as an optimal choice, since, from the engineering point of view, narrower channels could be very difficult to build, and maintain during the target operation. The length of the target itself is 5.28 cm. The first U-Mo plate is cooled from one side only, which is sufficient to keep the maximum temperature in this plate below 450 °C. The thickness of the inlet and outlet channels is 0.5 cm, while the length might change; it will depend on the configuration of the cooling system providing mercury coolant into the target.

From the neutronics point of view, the comparison of the initial compact design (ten 3 x 3 cm² U-Mo plates) and the final design reveals that no significant improvement could be obtained by additional optimization. An improvement of only 4.5% in the flux at 90° has been achieved in this case. It can be concluded that the new target proposed, as a final result of this thesis, improves significantly the angle-dependent flux, while maintaining the 4π neutron yield. For the angle of 72°, the neutron flux of the final design is higher by a factor of 12.1 at maximum. For the angle of 108°, the amount of material between the detector and the neutron-creation in the rotary target is much lower than for 72°. Therefore, at 108° the neutron flux of the final design is higher by a factor of 1.4 at maximum, compared with the rotary target. The final design also represents a more isotropic neutron source than the rotary target, thus users would profit from similar flux conditions at different flight paths.

A comparison of the resolution functions shows that the compact design provides better results than the current rotary target. The initial compact design and the final design produce comparable resolution functions, which again leads to the conclusion that no additional target optimization is required. Results also show that, in general, differences decrease with increasing energy, and no significant improvement of the resolution functions can be achieved above 3 MeV.

To perform a quantitative comparison of various designs, a figure of merit (FOM) was defined and used. For the full width at one half of the maximum (FWHM) values of the resolution functions, at angle of 72°, the FOM for the final design is higher by a factor of 20 at maximum, compared with the rotary target. At 108°, this decreases to a factor of 1.4. For FWTM values of the resolution functions these factors are 33 and 1.5 at 72° and 108°, respectively. This shows that, besides the improved flux at angles smaller than 108°, the main improvement is in the reduction of the tailing of the resolution functions. The

FOM comparison (for FWHM values) of the initial compact design and the final design shows improvements of 6.6%, 7.8%, and 7.8% at maximum for angles of 72°, 90°, and 108°, respectively.

The results presented in this thesis confirm that building the new target proposed would fulfil all project goals. Still, the next step, before the actual building of the target, should be the verification of the CFD simulations by means of temperature measurements on a mock-up model. Such a model can be heated artificially, which would prevent any material activation. This benchmark is necessary for security reasons, and for the official acceptance of the new design. Other investigations should be carried out to design a new moderator, which would enhance the performance of the new target in the low-energy range. It is concluded in this thesis that the low-energy range is directly related to the shape and composition of the moderator. This would further improve the neutronics properties of the GELINA facility for the low energy range, once the new target is in use.

A compact fast-neutron producing target for high resolution cross section measurements

A proper knowledge of neutron cross sections is very important for the operation safety of various nuclear facilities. Reducing uncertainties in the neutron cross sections can lead to an enhanced safety of present and future nuclear power systems. Accurate neutron cross sections also play a relevant role in many other disciplines such as astrophysics, medicine, and security. Therefore it is essential to have at disposal tools to measure the neutron cross sections at required resolution. The measurement accuracy required to extract properly resonance parameters of the cross sections can only be obtained at time-of-flight (TOF) facilities specially designed to have a high resolution in energy. Among the other neutron TOF facilities available in the world, the Geel electron linear accelerator (GELINA) facility of the Joint Research Centre of the European Commission is the one with the best energy resolution. The main goal of this thesis was to investigate the possibilities to improve even further the capabilities of this neutron data measurement facility. The present GELINA capabilities will be improved by designing a new high-power neutron producing target. The new target designed will substantially enhance the obtainable energy resolution, while not compromising the neutron flux.

It was realized at the very beginning of the project that the optimal size of a new target should be the result of a compromise between the neutron production and the quality of the resolution functions. The neutron production increases with the volume of the target. On the other hand, the quality of the resolution functions decreases with increasing volume. The MCNP results prove that a substantial reduction of the target volume, compared to the present GELINA target, will lead to improved resolution functions, while maintaining the overall neutron yield. The MCNP calculations show that the 4π neutron escape of the present target, which is about 5.6% n/e can be fully maintained by a U-Mo cylinder with R = 1.5 cm, and H = 5 cm. For that reason this cylinder was chosen as a starting geometry to design an improved target.

The first step towards a new target was to choose the best target material available. From the group of promising target materials, consisting of uranium, tantalum, tungsten, thorium, and mercury, uranium was chosen as a best material from the neutronics point of view. This material produces the highest amount of neutrons, compared to other materials tested. The 4π neutron yield is lower by 45%, if other material than uranium is used from the group mentioned above. The use of thorium would cause a 20% loss only. The different materials were compared using the targets scaled by quantities specifying an electromagnetic cascade. The cylindrical geometry with carefully chosen radius fully covers the electromagnetic cascade of high-energy gammas capable of neutron creation, so it is a reasonable choice for a target geometry. Moreover, such a compact target provides flight paths at different angles with similar neutron fluxes, and thus represents a much more isotropic neutron source than the actual rotary target. In the rotary target there is a large amount of material between the detector and the neutron-creation volume, which varies with the angle of a flight path. This heavy material scatters a large number of high-energy neutrons away from the track towards the detector. U-Mo alloy with 10%wt. of molybdenum was chosen instead of pure metallic uranium for the present rotary design to improve the mechanical properties of the target material. The presence of molybdenum has only a negligible impact on the neutronics properties of a compact U-Mo target, and therefore this alloy will be employed in a new design.

Based on these observations, a decision was made to focus on the high-energy range, since the target itself does not have any significant influence on the resolution functions for low neutron energies. The low-energy range is left for future investigation. The neutronics properties of the target-moderator assembly are mainly given by the moderator properties, and therefore a separate moderator design study will be performed in the future. As mercury performed well in the neutron yield simulations, this material was chosen as a coolant of a new target design. The use of heavy material for target cooling comes from the project conditions to produce high- and low-energy neutrons separately by using an assembly target-moderator. In this way the GELINA facility would not lose the flexibility of measurements, which nowadays can be performed simultaneously at many flight paths for a broad energy range.

As a second step of the investigation a decision was made to optimize a block-shaped geometry, instead of the cylindrical geometry. It was foreseen from the engineering point of view that such a design would be a more reasonable choice. Consequently, further optimization has been based on this type of design. The design of 3 x 3 x 5 cm^3 made of U-Mo produces a flux higher by 3.7%, compared to the U-Mo cylinder with H = 5 cm and R = 1.5 cm. This small difference is caused by additional material, which contributes to the neutron production.

Concerning the resolution functions, tantalum and tungsten are comparable with U-Mo. Both these materials could be reasonable substitutes for U-Mo to provide equivalent resolution functions. It was also realized that the resolution functions could not be improved above a neutron energy of 3 MeV. The improvement can be observed only in the tail of the resolution functions below the full width at one tenth of the maximum, which is of very low importance.

It is important to realize that the first phase change of uranium appears at 668 °C. However, at a nominal power level of 10 kW the maximum temperature of a single-piece U-Mo target would exceed this temperature, which would lead to target damage, caused by dimensional changes of the material. Therefore, a further investigation was aimed at optimizing a compact target with respect to reducing its maximum temperature by segmentation of the U-Mo block into right-angle plates perpendicular to the target axis. The channels in between the plates created by segmentation would allow mercury to flow through. For this reason an optimization was required to minimize the amount of Hg between the U-Mo plates, and to maximize the thickness of each U-Mo plate.

A temperature of 450 °C was chosen as a conservative upper limit to be respected by the design. A channel thickness of 1 mm has been chosen as an optimal choice, since, from the engineering point of view, narrower channels could be very difficult to build, and maintain during the target operation. The optimization was performed by positioning a U-Mo plate at the location of the highest neutron production. This location was found using the MCNP calculations for a solid U-Mo block of 3 x 3 x 5 cm³. For the same position also the power density along the axis was calculated by MCNP. Using this value the thickness of a plate was estimated such that its temperature did not exceed 450 °C. The same procedure was also applied to the other plates, while these were always separated by a 1-mm cooling channel. Afterwards, due to approximations employed in the analytical calculations, an additional correction on the thickness of some plates was required, as their maximal temperature exceeded the limit. This was realized from computational fluid dynamics simulations performed with the FLUENT code. Because the plates consist of radioactive material, a cladding was added to provide the required containment, and to minimize the contact of the U-Mo alloy with the coolant. Tantalum, as a good neutron producing material was chosen for the cladding. Finally, narrow inlet and outlet channels were added to the geometry. In this way, the final target design was created. The final design consists of seven 3 x 3 cm^2 U-Mo plates with thicknesses in the direction of the beam of 0.15 cm, 2 x 0.2 cm, 0.3 cm, 0.45 cm, 0.9 cm, and 2.1 cm. A tantalum cladding with a thickness of 0.2 mm coats each U-Mo plate. Seven 1-mm mercury channels located in between the plates cool the target. The length of the target itself is 5.28 cm. The first U-Mo plate is cooled only from one side, which is sufficient to keep the maximum temperature in this plate below 450 °C. The thickness of the inlet and outlet channels is 0.5 cm, while the length might change; it will depend on the configuration of the cooling system providing mercury coolant into the target.

It can be concluded that the new target proposed, as a final result of this thesis, improves significantly the angle-dependent flux, while maintaining the 4π neutron yield. For the angle of 72°, the neutron flux of the final design is higher by a factor of 12.1 at maximum. For the angle of 108°, the amount of material between the detector and the neutron-creation in the rotary target is much lower than for 72°. Therefore, at 108° the neutron flux of the final design is higher by a factor of 1.4 at maximum, compared with the rotary target.

A comparison of the resolution functions shows that the compact design provides better results than the current rotary target. Results show that differences decrease with increasing energy, and no significant improvement of the resolution functions can be achieved above 3 MeV.

To perform a quantitative comparison of various designs, a figure of merit (FOM) was defined and used. It is defined as the average neutron flux at the end of the flight distance necessary to obtain a given energy resolution. For FWHM values of the resolution functions, at angle of 72°, the FOM for the final design is higher by a factor of 20 at maximum, compared with the rotary target. At 108°, this decreases to a factor of 1.4. For FWTM values of the resolution functions these factors are 33 and 1.5 at 72° and 108°, respectively. This shows that, besides the improved flux at angles smaller than 108°, the main improvement is in the reduction of the tailing of the resolution functions.

Een compact snelle-neutronen producerende trefplaat voor hoge-resolutie metingen van werkzame doorsneden

Een gedegen kennis van werkzame doorsneden voor neutroneninteractie is belangrijk voor het bedrijven van diverse nucleaire installaties. De reductie van onzekerheden in deze werkzame doorsneden kan leiden tot verbeterde veiligheid van huidige en toekomstige nucleaire systemen. Nauwkeurige werkzame doorsneden spelen tevens een belangrijke rol in andere disciplines zoals astrofysica, geneeskunde en veiligheid (security). Het is daarom essentieel apparatuur beschikbaar te hebben met de benodigde resolutie ter meting van werkzame doorsneden. De benodigde nauwkeurigheid om resonantie parameters af te leiden kan alleen bij speciaal ontworpen zogenaamde time-of-flight (TOF) faciliteiten met een hoge energie resolutie worden bereikt. Van de beschikbare neutronen TOF faciliteiten heeft de Geel electron linear accelerator (GELINA) van het Joint Research Centre van de Europese Unie de beste energie resolutie. Het hoofddoel van dit onderzoek was om te onderzoeken of de mogelijkheden van deze faciliteit verder vergroot kunnen worden. De huidige mogelijkheden van GELINA kunnen worden verbeterd door een nieuwe hoog-vermogen trefplaat te ontwerpen. Deze nieuwe trefplaat zal de energie resolutie substantieel verbeteren dit met behoud van neutronenflux.

De optimale grootte van de trefplaat is een compromis tussen de neutronenproductie en de kwaliteit van de resolutie functies. De neutronenproductie neemt toe met het volume van de trefplaat. De kwaliteit van de resolutie functies neemt af met een toenemend volume. De MCNP resultaten laten zien dat een aanzienlijke reductie van het volume van de trefplaat mogelijk is en dat dit leidt tot verbeterde resolutiefuncties bij gelijkblijvende neutronen opbrengst. De MCNP resultaten laten zien dat de 4π neutronen opbrengst van de huidige trefplaat (ongeveer 5.6% n/e) ook behaald kan worden met een trefplaat in de vorm van een cilinder met R = 1.5 cm en H = 5 cm. Een dergelijke cilinder is als uitgangspunt genomen voor een verbeterd ontwerp.

De eerste stap naar een nieuwe trefplaat was om het beste materiaal te kiezen. Uit de serie van meest veelbelovende materialen, te weten uranium, tantallum, wolfraam, thorium en kwik is uranium de beste keus bezien vanuit de neutronica. Dit materiaal produceert namelijk het grootste aantal neutronen. De opbrengst verlaagt met 45% als een ander materiaal wordt gekozen. Een uitzondering betreft thorium waar de vermindering slechts 20% bedraagt. De verschillende materialen zijn vergeleken in trefplaten die geschaald waren volgens de maat van de elektromagnetische cascade in het materiaal. Een cilindrische geometrie met een zorgvuldig gekozen radius bedekt volledig de elektromagnetische cascade van hoge-energie gamma's, die op hun beurt neutronen kunnen losmaken. Verder leidt een dergelijk compacte geometrie tot een vergelijkbare neutronenflux in de diverse vluchtpaden, m.a.w. de bron is meer isotroop dan de huidige roterende trefplaat. In de huidige trefplaat is, afhankelijk van de vluchtpadhoek, een grote hoeveelheid materiaal tussen de detector en het volume waar de neutronen geproduceerd worden. Het zware materiaal van de trefplaat verstrooit een groot deel van de neutronen en verlaagt effectief de neutronenopbrengst in de richtingen van de detector. Een legering van U-Mo is gekozen als materiaal voor de nieuwe trefplaat vanwege de betere mechanische eigenschappen dan van puur uranium. De aanwezigheid van molybdeen heeft slechts een kleine invloed of de neutronenopbrengst van het compacte U-Mo ontwerp, daarom is deze legering gekozen als basis voor de rest van deze studie.

Op basis van deze resultaten is besloten om op het hoge energie gebied te concentreren omdat de trefplaat zelf een kleine invloed heeft op de resolutie functies in het lage-energie gebied. Het laatst genoemde gebied zal in een andere studie bekeken worden. Daar de neutronica eigenschappen van de trefplaat-moderator combinatie worden voornamelijk bepaald door de moderator, zal een aparte studie in de toekomst gewijd worden aan het ontwerp van een nieuwe moderator. Kwik, het koelmiddel van de bestaande trefplaat, is wegens de goede resultaten van de simulaties ook gekozen als koelmiddel van de nieuwe trefplaat. Het gebruik van een koelmiddel met een hoog atoomgetal komt van de eis dat hoge- en lage-energie neutronen afzonderlijk geproduceerd worden in een trefplaat-moderator combinatie. Op deze manier kan de GELINA faciliteit de flexibiliteit behouden waarbij metingen kunnen worden verricht in verschillende vluchtpaden voor een breed energiespectrum.

Als tweede stap is de beslissing genomen om een blokvormige geometrie te bestuderen i.p.v. een cilindrische geometrie, dit vanuit het oogpunt van simpeler constructie. Verdere optimalisatie is dan ook gebaseerd op een dergelijke geometrie. Het ontwerp met een afmeting van 3 x 3 x 5 cm^3 van U-Mo levert een neutronen opbrengst van 3.7% groter dan een cilinder met H = 5 cm en R = 1.5 cm van hetzelfde materiaal. Dit kleine verschil wordt verklaard door het additionele materiaal dat in kleine mate nog bijdraagt aan de productie van neutronen.

Op het gebied van de kwaliteit van de resolutie functies zijn tantalium en wolfraam vergelijkbaar met U-Mo. Tevens kunnen de resolutie functies boven 3 MeV nauwelijks verbeterd worden; de verbeteringen bevinden zich slechts in de staart van de resolutie functies beneden de grens van de "full width at one tenth of the maximum", waar verbeteringen een geringe invloed hebben.

De eerste fase overgang van U-Mo vindt plaats bij 668 °C. Bij gebruik van een trefplaat gemaakt uit een massief blok U-Mo en een gedeponeerd vermogen van 10 kW zou deze temperatuur overschreden worden. Op zijn beurt zou dit leiden tot schade in de trefplaat. Het verdere onderzoek was dan ook gericht op de optimalisatie van een compacte trefplaat zodanig dat de maximum temperatuur verlaagd wordt door het blok op te delen in verschillende rechthoekige platen. De kanalen tussen de platen kunnen dan gebruikt worden om kwik te laten doorstromen. Hierom is een optimalisatie vereist om de hoeveelheid kwik tussen de U-Mo platen te minimaliseren en de dikte van de platen zelf te maximaliseren.

Een temperatuur van 450 °C is gekozen als conservatieve limiet waar het ontwerp aan dient te voldoen. Een kanaaldikte van 1 mm is gekozen omdat een trefplaat met smallere kanalen waarschijnlijk moeilijk te construeren en onderhouden zou zijn. Een optimalisatie studie is gestart waarbij een plaat gepositioneerd wordt op de locatie van de maximale neutronenproductie. Deze locatie is bepaald d.m.v. MCNP berekeningen voor een massief blok U-Mo van 3 x 3 x 5 cm³. De vermogensdichtheid op deze locatie kan ook worden berekend met MCNP. Met deze waarde kan op basis van een analytische schatting de dikte van de plaat bepaald worden zodanig dat de maximale temperatuur niet overschreden wordt. Een vergelijkbare procedure is toegepast op de andere platen waarbij altijd een 1 mm koelkanaal tussen de platen is geplaatst. Achteraf is een correctie toegepast op de diktes van de platen omdat de maximale temperaturen in FLUENT simulaties toch de limiet overschreden. Dit is het gevolg van de eerder genoemde analytische benaderingen. Omdat de platen na bestraling radioactief materiaal bevatten, is een cladding nodig om een omhulsel te vormen en om contact met het koelmiddel zoveel mogelijk te vermijden. Tantalium, is met een goede neutronen opbrengst, gekozen als cladding materiaal. Tenslotte zijn smalle in- en uitlaat secties toegevoegd aan de geometrie voor de aan- en afvoer van het koelmiddel. Het uiteindelijke ontwerp bestaat uit zeven 3 x 3 cm² platen met diktes van 0.15 cm, 2 x 0.2 cm, 0.3 cm, 0.45 cm, 0.9 cm, en 2.1 cm. Een tantalium cladding met een dikte van 0.2 mm bedekt iedere U-Mo plaat. Zeven 1 mm koelkanalen tussen de platen koelen de trefplaat. De eerste plaat is slechts gekoeld van een kant wat voldoende blijkt te zijn om de temperatuur beneden 450 °C te houden. De dikte van de aan- en afvoer kanalen is 0.5 cm en de lengte kan variëren afhankelijk van de uiteindelijke configuratie benodigd voor koelmiddel aan- en afvoer.

De in dit proefschrift voorgestelde nieuwe trefplaat geeft een aanzienlijke verbetering van de hoekafhankelijkheid van de flux terwijl de totale opbrengst min of meer gelijk blijft. Voor een hoek van 72° de neutronenflux is 12.1 keer groter dan in de bestaande roterende trefplaat. Bij een hoek van 108° is de hoeveelheid materiaal tussen de locatie van neutronenproductie en detector veel kleiner en daarom is de neutronenflux in het nieuwe ontwerp slechts 1.4 keer groter dan in het huidige ontwerp.

Een vergelijk tussen de resolutie functies van de huidige roterende plaat en het nieuwe compacte ontwerp laat zien dat de laatste veel beter van kwaliteit zijn. De verschil wordt kleiner bij hogere energieën en geen significante verbetering kan worden bereikt boven 3 MeV.

Om tot een kwantitatief vergelijk tussen verschillende ontwerpen te komen is een "Figure Of Merit (FOM)" gedefinieerd en toegepast. Dit FOM is gedefinieerd als de gemiddelde neutronenflux aan het eind van een vluchtpad benodigd voor een gegeven energie resolutie. Voor "Full Width Half Maximum (FWHM)" waarden van de resolutie functies, bij een hoek van 72°, is de FOM van het uiteindelijke ontwerp een factor 20 hoger op het maximum dan bij de bestaande roterende trefplaat. Bij 108° verlaagt dit tot een factor 1.4. Voor "Full Width Tenth Maximum (FWTM)" waarden van de resolutiefuncties zijn de factoren 33 en 1.5 bij respectievelijk 72° en 108°. Dit laat zien dat, afgezien van de verbeteringen in flux voor hoeken lager dan 108°, de belangrijkste verbetering in ligt in reductie van de staart van de resolutie functies.

References

[Bar96]   R.M. Barnett et al., Particle Data Group, "Part I - Review of particle physics," *Phys. Rev.*, **D54**, 132, (1996).

[Bat70]   G. Bathow, E. Freytag, M. Kobberling, K. Tesch, R. Kajikawa, "Measurements of the Longitudinal and Lateral Development of Electromagnetic Cascades in Lead, Copper and Aluminum at 6 GeV," *Nucl. Phys.*, **B20**, 592, (1970).

[Bee00]   W.J. Beek, K.M.K. Muttzall, J.W. van Heuven, *"Transport Phenomena,"* John Wiley & Sons Ltd, Great Britain (2000).

[Ben78]   A. Bensussan, J.M. Salome, "Gelina: a modern accelerator for high-resolution neutron time of flight experiments," *Nucl. Inst. and Meth.*, **155**, 11–23, (1978).

[Bir02]   R.B. Bird, W.E. Stewart, E.N. Lightfoot, *"Transport Phenomena,"* John Wiley & Sons Inc, USA (2002).

[Bir03]   T. Biro, "Neutrons as tools in safeguards and combating illicit trafficking of nuclear material," *Proceedings of the workshop The Nuclear Measurements and Evaluations for Applications (NEMEA-2)*, Budapest, Hungary (November 5-8, 2003).

[Bor05]   A. Borella, *"Determination of the neutron resonance parameters for ^{206}Pb and of the thermal neutron capture cross section for ^{206}Pb and ^{209}Bi,"* Ph.D. thesis, Ghent University, Belgium (2005).

[Brie00]   J.F. Briesmeister, ed., *"MCNP - A General Monte Carlo N-Particle Transport Code, Version 4C,"* LA-13709-M, LANL, Los Alamos, USA (April 2000).

[Bru02]   A. Brusegan, G. Noguere, F. Gunsing, "The resolution function in neutron time-of-flight measurements," *J. Nucl. Sci. and Techn.*, **Suppl. 2**, 685, (2002).

[Cal80a]   J.T. Caldwell, E.J. Dowdy, B.L. Berman, R.A. Alvarez, P. Meyer, "Giant resonance for actinide nuclei: photoneutron and photofission cross sections for ^{235}U, ^{236}U, ^{238}U and ^{232}Th," *Phys. Rev.*, **C21**, 1215, (1980).

[Cal80b]   J.T. Caldwell, E.J. Dowdy, R.A. Alvarez, B.L. Berman, P. Meyer, "Experimental determination of photofission neutron multiplicities for ^{235}U, ^{236}U, ^{238}U, ^{232}Th using monoenergetic photons," *Nucl.Sci.and Eng.*, **73**, 153, (1980).

[Coc02]   C. Coceva, M. Frisoni, M. Magnani, A. Mengoni, "On the figure of merit in neutron time-of-flight measurements," *Nucl. Inst. and Meth. A*, **489**, 346-356, (2002).

[Coc83]   C. Coceva, R. Simonini, D.K. Olsen, "Calculation of the ORELA neutron moderator spectrum and resolution function," *Nucl. Inst. and Meth.*, **211**, 459-467, (1983).

[Coc96]   C. Coceva, M. Magnani, *"Resolution Rotary Target,"* Internal report GE/R/ND/06/96, EC-JRC-IRMM, Geel, Belgium (1996).

[Cor02]   F. Corvi, G. Fioni, F. Gunsing, P. Mutti, L. Zanini, "Resonance Neutron Capture in ^{60}Ni below 450 keV," *Nucl. Phys. A*, **697**, 581-610, (2002).

[Die88]   S.S. Dietrich, B.L. Berman, "Atlas of photoneutron cross sections obtained with monoenergetic photons," *Atomic Data and Nuclear Data Tables*, **38**, 199-338, (1988).

[Eva55]   R.D. Evans, *"The atomic nucleus,"* McGraw-Hill Book Company, Inc., USA (1955).

[Fla03]   M. Flaska, D. Lathouwers, T.H.J.J. van der Hagen, H. van Dam, W. Mondelaers, A.J.M. Plompen, A. Borella, P. Schillebeeckx, P. Rullhusen, "Study of properties of the GELINA neutron target," *Proceedings of The Nucl. Mathematic. And Comput. Sciences, A Century In Review – A Century Anew*, Gatlinburg, Tennessee, USA (2003).

[Fla04]   M. Flaska, A. Borella, D. Lathouwers, L.C. Mihailescu, W. Mondelaers, A.J.M. Plompen, H. van Dam, T.H.J.J. van der Hagen, "Modeling of the GELINA neutron target using coupled electron-photon-neutron transport with the MCNP4C3 code," *Nucl. Inst. and Meth. A*, **531**, 392-406, (2004).

[Fla05]   M. Flaska, D. Lathouwers, A.J.M. Plompen, W. Mondelaers, T.H.J.J. van der Hagen, H. van Dam, "Potential for improvement of a neutron producing target for time-of-flight measurements," *Nucl. Inst. and Meth. A*, **555**, 329-339, (2005).

[Flu01]   FLUENT 6 user's guide, Lebanon, New Hampshire, USA (2001).

[Fro94]   T. Frommhold, F. Steiper, W. Henkel, U. Kneissl, J. Ahrens, R. Beck, J. Peise, M. Schmitz, I. Anthony, J.D. Kellie, S.J. Hall, G.J. Miller, "Photofission of U-235 and U-238 at intermediate energies: Absolute cross-sections and fragment mass distributions," *Z. Phys.*, **A350**, 249-261, (1994).

[Gri89]   G. Grindhammer et al., *Proceedings of the Workshop on Calorimetry for the Supercollider*, Tuscaloosa, Alabama, March 13-17, 1989, edited by R. Donaldson and M.G.D. Gilchriese, 151, World Scientific, Teaneck, New York, USA (1989).

[Gro47]   H.J. Groenewold, H. Groendijk, *Physica XIII*, **1-3**, 141-152, (March 1947).

[Gun00]   F. Gunsing, A. Lepretre, C. Mounier, C. Raepsaet, A. Brusegan, E. Macavero, "Neutron Resonance Spectroscopy of ^{99}Tc from 3 eV to 150 keV," *Phys. Rev. C*, **61**, 054608, (2000).

[IAE00]   IAEA-TECDOC-1178, "*Handbook on photonuclear data for applications: Cross sections and spectra,*" Vienna, Austria (2000).

[Kor79]   I.S. Koretskaya, V.L. Kuznetsov, L.E. Lazareva, V.G. Nedoresov, N.V. Nikitina, *Yadernaja Fyzika*, **30**, 910, (1979).

[Lep87]   A. Lepretre, R. Bergere, P. Bourgeois, P. Carlos, J. Fagot, J.L. Fallou, P. Garganne, A. Veyssiere, H. Ries, R. Goeble, U. Kneissl, G. Mank, H. Stroeher, W. Wilke, D. Ryckbosch, J. Jury, "Absolute photofission cross sections for ^{232}Th and ^{235,238}U measured with monochromatic tagged photons (20 Mev $< E_\gamma < $ 110 Mev)," *Nucl. Phys.*, **A472**, 533, (1987).

[Lil01]   J. Lilley, "*Nuclear Physics – Principles and Applications,*" John Wiley & Sons Ltd, Great Britain (2001).

[Mih04]   L.C. Mihailescu, L. Olah, C. Borcea, A.J.M. Plompen, "A new HPGe setup at GELINA for measurement of gamma-ray production cross-sections from inelastic neutron scattering," *Nucl. Inst. and Meth. A*, **531**, 375-391, (2004).

[Mox89]   M. Moxon, "*REFIT2: A least Squares Fitting Program for Resonance Analysis of Neutron Transmission and capture Data,*" NEA-0914/02 (1989).

[Nel66]   W.R. Nelson, T.M. Jenkins, R.C. McCall, J.K. Cobb, "Electron-Induced Cascade Showers in Copper and Lead at 1 GeV," *Phys. Rev.*, **149**, 201, (1966).

[Ost78]   Y.B. Ostapenko, G.N. Smirenkin, A.S. Soldatov, V.E. Zhuchko, Y.M. Tsipenyuk, *Yadernaja Konstanti-3*, **30**, 3, (1978).

[Per00]   D.H. Perkins, "*Introduction to High Energy Physics,*" Cambridge University Press, Cambridge, Great Britain (2000).

[Qai02]   S.M. Qaim, "Nuclear data for production of new medical radionuclides," *J. Nucl. Sci. and Techn.*, **Suppl. 2**, 1272, (2002).

[Rae70]   E.R. Rae, W.M. Good, in: J.A. Harvey, ed., *"Experimental Neutron Resonance Spectroscopy,"* Academic Press, New York, USA (1970).

[Rie84]   H. Ries, G. Mank, J. Drexler, R. Heil, K. Huber, U. Kneissl, R. Ratzek, H. Stroeher, T. Weber, W. Wilke, "Absolute photofission cross sections for 235,238U in the energy range 11.5-30 MeV," *Phys. Rev.*, **C29**, 2346-2348, (1984).

[Ros52]   B. Rossi, *"High Energy Particles,"* Prentice-Hall Inc., New York, USA (1952).

[Ryc88]   D. Ryckbosch, P. Carlos, A. Lepretre, "Hybrid model analysis of intermediate energy photonuclear reactions on 232Th and 235,238U," *Z. Phys.*, **A329**, 451, (1988).

[Sal02]   M. Salvatores, "Future nuclear power systems and nuclear data needs," *J. Nucl. Sci. and Techn.*, **Suppl. 2**, 4, (2002).

[Sal81]   J.M. Salome, R. Cools, "Neutron producing target at GELINA," *Nucl. Inst. and Meth.*, **179**, 13-19, (1981).

[Smi02]   M.S. Smith, "Nuclear data relevant to astrophysics," *J. Nucl. Sci. and Techn.*, **Suppl. 2**, 19, (2002).

[Sta01]   W.M. Stacey, *"Nuclear reactor physics,"* John Wiley and Sons, Inc., ISBN 0-471-39127-1, New York, USA (2001).

[Tro85]   D. Tronc, J.M. Salome, K. Böckhoff, "A new pulse compression system for intense relativistic electron beams," *Nucl. Inst. and Meth.*, **228**, 217-227, (1985).

[Vey73]   A. Veyssiere, H. Beil, R. Bergere, P. Carlos, A. Lepretre, "A study of the photofission and photoneutron processes in the giant dipole resonance of ^{232}Th, ^{238}U and ^{237}Np," *Nucl. Phys.*, **A199**, 45, (1973).

[Whi00]   M.C. White, *"Release of the LA150U Photonuclear Data Library,"* X-5:MCW-00-87(U), LANL, Los Alamos, USA (2000).

[Zie00]   O.C. Zienkiewicz, R.L. Taylor, *"The finite element method,"* Butterworth-Heinemann, USA (2000).

List of publications

1. M. Flaska, D. Lathouwers, T.H.J.J. van der Hagen, H. van Dam, W. Mondelaers, A.J.M. Plompen, A. Borella, P. Schillebeeckx, P. Rullhusen, "Study of properties of the GELINA neutron target," *Proceedings of The Nucl. Mathematic. And Comput. Sciences, A Century In Review – A Century Anew,* Gatlinburg, Tennessee, USA (2003).

2. M. Flaska, A. Borella, D. Lathouwers, C. Mihailescu, W. Mondelaers, A.J.M. Plompen, T.H.J.J. van der Hagen, H. van Dam, "Optimization Study of the GELINA Neutron Target," *Proceedings of The Nuclear Measurements and Evaluations for Applications (NEMEA),* Budapest, Hungary (2003).

3. M. Flaska, A. Borella, D. Lathouwers, L. C. Mihailescu, W. Mondelaers, A.J.M. Plompen, H. van Dam, T.H.J.J. van der Hagen, "Towards an Improved GELINA Neutron Target," *Proceedings of The Physics of Fuel Cycles and Advanced Nuclear Systems: Global Developments (PHYSOR 2004),* Chicago, USA (2004).

4. M. Flaska, A.J.M. Plompen, W. Mondelaers, D. Lathouwers, T.H.J.J. van der Hagen, H. van Dam, "GELINA Neutron Target Optimisation," *Proceedings of The International Conference on Radiation Shielding (ICRS-10), the topical meeting on Radiation Protection and Shielding (RPS 2004),* Madeira, Portugal (2004).

5. M. Flaska, A. Borella, D. Lathouwers, L.C. Mihailescu, W. Mondelaers, A.J.M. Plompen, H. van Dam, T.H.J.J. van der Hagen, "Modeling of the GELINA neutron target using coupled electron-photon-neutron transport with the MCNP4C3 code," *Nucl. Inst. and Meth. A,* **531**, 392-406, (2004).

6. M. Flaska, D. Lathouwers, A.J.M. Plompen, W. Mondelaers, T.H.J.J. van der Hagen, H. van Dam, "New GELINA Target Development – project status," *Proceedings of The Nuclear Measurements and Evaluations for Applications (NEMEA-2),* Bucharest, Romania (2004).

7. M. Flaska, A.J.M. Plompen, W. Mondelaers, D. Lathouwers, T.H.J.J. van der Hagen, H. van Dam, C.R. Kleijn, "Modeling of an Improved GELINA Target for the Production of Neutrons," *Proceedings of Mathematics and Computation, Supercomputing, Reactor Physics and Nuclear and Biological Applications (M&C 2005)*, Avignon, France (2005).

8. M. Flaska, D. Lathouwers, A.J.M. Plompen, W. Mondelaers, T.H.J.J. van der Hagen, H. van Dam, "Potential for improvement of a neutron producing target for time-of-flight measurements," *Nucl. Inst. and Meth. A*, **555**, 329-339, (2005).

Acknowledgements

The research described in this thesis was carried out in cooperation between the TU Delft, The Netherlands, and the Institute for Reference Materials and Measurements (IRMM) of the Joint Research Centre of the European Commission in Geel, Belgium. Therefore I would like to express my acknowledgements to people from both organizations.

First of all, I would like to express my great gratitude to Prof. van der Hagen and Prof. van Dam who, as my promotors, gave me the opportunity to work at IRI. Together with the other promotor, Prof. Kleijn, they regularly provided me during our meetings with highly scientific and always-to-the-point comments and suggestions. Their positive criticism allowed me to learn the way of "scientific thinking". I am very grateful for their rigorous supervision, which minimized my inborn tendency to explore research problems, which are not relevant to the main task, but still very interesting and attractive. This helped me to "stay on the track" and finish the job within four years, as initially agreed. Secondly, I would like to thank Danny Lathouwers who, as my daily IRI supervisor during the initial year of my project, significantly helped me to start up the project. His essential contributions originated from our intensive and long scientific discussions, which always ended up with fruitful ideas and conclusions. Even during the remaining three years of the project, during which I was working as a fellow at the IRMM, I could always rely on Danny's support, not only for what concerned professional matters, but also concerning private and social issues. Here I also want to thank to Piet de Leege for his ability to solve any software and(or) data problem, usually in a surprisingly short time.

As my IRMM fellowship was founded by European Commision (EC), at this place I would like to thank the EC officials for the opportunity to work at IRMM. Starting with IRMM people, I express my thanks to Prof. Rullhusen for the chance to be involved in the GELINA project, and to be employed by the IRMM at the Neutron Physics Unit. My great gratitude is given to Arjan Plompen and Wim Mondelaers, who acted as my daily IRMM supervisors. Their persisting effort for a scientific dialog significantly improved the quality of the output of my work. I appreciate very much all discussions we had, because our meetings were always very productive and helpful. I could profit many times from their professional experience, and I am very thankful for that. In addition, I could always count on their support, even if the issue was not of a professional kind. Next I express my acknowledgements to Peter Siegler and Peter Schillebeeckx, whose professional experience in the field of neutron cross section measurements facilitated my understanding of the project objectives for designing an improved GELINA target. Moreover, thanks to them and their families, my wife and me felt like at home in Belgium. I also thank to Alessandro Borella and Cristian Mihailescu for providing me with the neutron flux measurement data, and for the explanations of their experimental setups.

During my PhD research I was living in two countries. During both stays I met many very friendly and sociable people, and many of them became my good friends. Thanks to them, my life abroad was always very pleasant and enjoyable. Many thanks to all of you!

Finally, these acknowledgements would not be complete without mentioning the person who is essential to my life; that is my wife. Thank you Maria for your love, for staying besides me at all times, and for your talent to bring my (sometimes very) negative moods back to the positive side. I also want to express my thanks to other family members for their moral support; this work would not be done without you.

Marek Flaška
Born on 5th of January 1977 in Veľký Krtíš, Slovak Republic

1991-1995 — Jozef Murgaš Secondary Electrotechnical School, Banská Bystrica, Slovak Republic

1995-2001 — Master degree, Slovak University of Technology, Faculty of Electrical Engineering and Information Technology, Bratislava, Slovak Republic
Graduation thesis: *The influence of density-optimizing additives on UO_2 pellet properties*
Baccalaureate thesis: *The leakage calculations from the spent-fuel transport container C-30*

2001-2005 — PhD student of the Faculty of Applied Sciences of the Delft University of Technology, Department of Radiation, Radionuclides and Reactors, Delft, The Netherlands

2002-2005 — PhD-related research performed at Neutron Physics Unit of the Institute for Reference Materials and Measurements in Geel, Belgium.